The Ocean World of Jacques Cousteau

Provinces of the Sea

The Ocean World of Jacques Cousteau

Volume 11

Provinces of the Sea

THE DANBURY PRESS

A burning sun plays its golden light, revealing only a pacific face that hides the ocean's layers of life, the majestic volcanic crags, the widemouth gashes of hadal trenches, and the ceaseless activity that characterizes all the provinces of the sea.

The Danbury Press
A Division of Grolier Enterprises Inc.

Publisher: Robert B. Clarke

Production Supervision: William Frampton

Published by The World Publishing Company

Published simultaneously in Canada
by Nelson, Foster & Scott Ltd.

ISBN O-529-05078-1
Library of Congress catalog card number: 73-11895

Printed in the United States of America

23456789987654

Project Director: Steven Schepp

Managing Editor: Ruth Dugan
Assistant Managing Editor: Christine Names
Senior Editor: David Schulz

Assistant Editor: Jill Fairchild
Editorial Assistant: Joanne Cozzi

Art Director and Designer: Gail Ash

Assistant to the Art Director: Martina Franz
Illustrations Editor: Howard Koslow

Production Manager: Bernard Kass

Science Consultants: Richard C. Murphy
                    Dr. David Schwimmer

Creative Consultant: Milton Charles

Typography: Nu-Type Service, Inc.

Contents

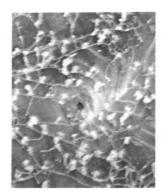

As the land slopes toward the sea, sometimes gently, sometimes precipitously, the true ends of the continents are hidden BELOW THE DEEPEST TIDES (Chapter VI). It is in the waters of this area where some of the most abundant life-producing activity takes place, both in the amount of individuals and the number of species.

The surface of the sea may seem to be only a broad, monotonous, almost never ending expanse of blue water broken occasionally by a leaping manta ray or a sounding whale. But there are times when THE OPEN OCEAN (Chapter VII) is reddened by dinoflagellates, giving a hint of the free-floating existence of so many plants and animals. And below the surface, there are multitudes of free-swimming creatures which range far and wide in pursuit of food and suitable waters.

FACING THE ABYSS (Chapter VIII), we find it difficult to imagine a world in which there is no light, constant cold, tremendous pressure and no green plant life. However some life accepts these conditions and makes a home in this harsh environment. Hidden in the abyss are peculiar topographic features like seamounts, guyots, and submarine canyons, while upon the bottom lie terrestrial sediments and biogenous oozes.

The least known of the undersea regions is THE HADAL ZONE (Chapter IX), with its deep trenches and bizarre forms of life existing at depths of 30,000 feet. An important part of this area, though, exists above the surface. These are the volcanoes which occur on island arcs bordering the trenches and believed to be integrally connected with them.

The origin and creation of the oceans, the processes by which they were formed, the separation of land and sea all border on a PROVINCE OF THE UNKNOWN (Chapter X). The effects of ocean changes can be observed, but the causes are hidden, perhaps so deep in the earth that they challenge the reach of man. The mechanism for ocean building can only be guessed at, even as new oceans may be in the process of being created.

By what manner and method did the oceans come to be where they are today? Or were they always there? Are the forces responsible for SHAPING THE EARTH (Chapter XI) still active today? The answers are as hard to come by as the questions are easy to phrase. The evidence is often contradictory, pointing first toward one solution, then to another.

The history of the continents, and mainly the history of life on our planet is, in fact, a WATER STORY.

Introduction: Our Water Planet

The continents, as well as the oceans are in constant evolution, in perpetual motion. The seabeds move, the landmasses collide, peninsulas sink, islands rise from the sea. Oceans expand and contract, continents grow and split. Entire life systems are born or wiped out.

The oceans encompass *hydros* or seawater itself, *lithos* or ocean floor, and *bios,* the life of the sea. The water that bathes the shores and teems with all the creatures it begat is in perpetual motion, up and down, back and forth, mixing, stirring, welling up, and sinking. When this action is rough, it creates an area of turbulence; when it is slow, it permits waters to settle in density layers, according to temperature and salinity. Well-defined currents, like the swift moving Gulf Stream, cut across and through the oceans.

There are distinct realms within the sea, provinces or zones which have their own characteristics, scenery, and populations. Such areas cannot be defined with absolute precision; the provinces are merely convenient concepts to help our alien minds perceive the majesty and dominion of the sea. We speak of the abyss or continental margins as though we were speaking of France or California with their well-marked boundaries. It is more nearly akin to speaking of the Orient or the Tropics. When one is there, one knows it. But where it starts, where it ends, becomes arbitrary.

The provinces of the sea, as everchanging as they are, provide the opportunity to study the spectacle of life, as though water were the medium, life its message. The provinces may be familiar, so familiar that they lose their mystery. Or they may be overly romanticized because of their remoteness. But life resists arbitrary boundaries, distributing itself according to the conditions of the environment rather than to man's measure of geography. True, not all animals are found everywhere, but the range of closely related species attests to the ability of organisms to adapt and survive. Whether it is the tiny life forms of the plankton—some plant, some animal, and some we are not so sure about—or the highly selective coral, the impression is one of a world of evolution and change, of profusion and fraility.

The vastness of the oceans, the immense range of waters, is what impressed the early man most. Since he started his difficult career in a hostile world, he wondered—as no other animal ever did—about stars, about thunder and storms, about the immense and impenetrable ocean. Then, as civilization upon civilization developed and replaced each other, the sea remained the obstacle that divided the world. From antiquity to the modern times of the great navigators, the sea was only used as a fishing ground, a highway for coastal trade, and a battlefield. The Americas were first populated during the last glaciation period, about 12,000 years ago, when the sea level was low enough to permit Asians to walk to Alaska and progressively invade the south and the east.

It is only very recently that the oceans no longer are a serious barrier to communication between civilizations. Philosophical concepts were deeply influenced by man's late awareness that all human beings were citizens of the same water planet. Classic humanities have flourished on the assumption that the birth of consciousness in man was the ultimate development of the mind. But the new philosophers think otherwise. Individual consciousness produced a wealth of creative masterpieces in liberal arts but ended up in a junglelike overemphasis on personal drives and produced an unchecked explosion of technology purely aimed at the artificial creation of individual needs. The process started in the eighteenth century with the steam engine was bolstered by publicity and drives us to the poisoning of air and water, to the eradication of hundreds of species . . . and to the assassination of the oceans, with all the consequences for mankind.

Fortunately, thanks to the space programs, a new kind of consciousness is developing in the people's mind: a global consciousness. Space exploration started with the ambitious dream of conquering outer space, landing on the moon, and later on Mars. This great project was a tremendous luxury when a third of mankind was still starving, when 15 percent of the population of the richest countries was still in a state of undignified poverty, and when 80 percent of mankind did not have adequate medical care. But space exploration brought back an unexpected, most precious, and most timely gift: a cosmic view of our earth. The photos of the earth materialized and popularized the concept that we were all passengers of the same small, but exceptionally rich and beautiful spaceship. Frontiers and borderlines that show so well on color maps, and for which so many men have died, were not visible on such pictures. There were oceans, continents, lakes, and rivers as a common heritage. There was no replacement for it in the solar system. We obviously had to become earth-conscious, water-conscious, energy-conscious, if we were to survive our population explosion.

From space it will be soon possible to assess the productivity of the oceans; to control forest fires, droughts, and floods; to measure the drift of continents and the height of the swell; to delineate fluctuations of currents; to prevent hurricanes; to pinpoint pollutions; to follow the tracks of whales or other migrating animals; and to interrogate thousands of instrumented oceanic buoys. Monitoring the oceans from manned space laboratories is the only hope we have that the looting of the sea will end and its rational management will begin.

Jacques-Yves Cousteau

Chapter I. Water of Life

Life originated in the ocean about three billion years ago, and the sea today remains a fountain of life, producing myriads of individuals in hundreds of thousands of different species each year. The chemical and physical properties of seawater made life possible. These properties, along with energy from the sun, provided the foundation upon which organic matter developed and gained the ability to perpetuate the species by repro-

---

"Water's ability to store large amounts of heat, and to release it slowly, helps the oceans moderate the earth's climate."

---

ductive processes. The indispensable energy comes, ultimately, from the sun; life in the upper reaches of the ocean receive this energy fairly directly, and the creatures below get it "secondhand." The nature of seawater is such that it continues to nourish life, from the microscopic plankton to the large ocean mammals and even to the air-breathers who live upon the continents.

The unique properties of water start with its magic molecule—a simple chemical compound consisting of two hydrogen atoms and a single oxygen atom. This molecule has a slight negative electrical charge at one end, the oxygen end, and a slight positive charge at the other, the hydrogen end. Because positives attract negatives, water then has a tendency to stick to itself and to a host of other chemical compounds, giving it an array of exceptional qualities.

This self-adhesive force is evident in "surface tension," which in water exceeds that of any other liquid. The high boiling point of water is another manifestation of this property. Water's ability to hold large amounts of heat—and to lose it very slowly—is what helps the oceans moderate the earth's climate.

Because the molecular structure of water prevents it from becoming easily separated by heat (it does not easily vaporize), we use water to cool ourselves, whether it is by sweating, by turning on air conditioners, or by going to the beach on a hot day.

Its molecular structure is the basis for all of water's amazing properties. These include water's ability to invade substances and carry away molecules, thus dissolving them, as it does salt. The combination of dissolved salt and temperature is responsible for water circulation, since the colder, denser water sinks and the warmer, lighter water rises. Saltier water also sinks and thus is established the mechanism behind ocean circulation, the constant flow that brings nutrients to live creatures. Water's fluidity distributes the dissolved salts, minerals, and other life elements throughout the ocean world and also throughout each organism, as the water circulates within it.

Seawater can dissolve gases—oxygen and carbon dioxide among others—which are integral parts of the life cycle. The development of marine plants produced the free oxygen that led to the composition of our present atmosphere. These plants, in turn, provide food for the animals. Thus a cycle is formed. Once closed, the circle must remain unbroken or disaster will result for all concerned.

The magic molecule. In exploring the life of the ocean, man has better assessed seawater's unique abilities to dissolve chemicals and gases, absorb great amounts of heat, and allow life-giving sunlight to penetrate the surface.

11

Salt of the Earth

Imagine the earth's surface being smoothed out and seawater covering the entire face of the planet. Then, in this make-believe world, imagine all the water evaporating. We would then be left with a crust of salt about 200 feet thick encasing the entire globe.

This salt is mainly sodium chloride, common table salt, but there are many other chemicals in the sea. In fact most, if not all, of the elements found on earth can be found in the oceans; the most common are sodium, potassium, magnesium, calcium, chlorine, bromine, and sulphur. Hydrogen and oxygen, of course, make up the water itself. The term salt, when referring to minerals, means a combination of elements, such as the first four listed above, with acidic elements, such as chlorine, bromine, and sulphur, which are usually combined with oxygen.

When the oceans were formed, they were bodies of fresh water. As weathering of rocks on land took place, salts were dissolved, ran

The materials entering the ocean, in addition to those brought by rivers, might come by way of windblown dust, airborne particles brought down by rainfall, meteorites from outer space, or volcanic dust. Offsetting these additions are the processes which withdraw elements from the sea. The most observable of these is sea spray, which when it is blown on to land areas evaporates and leaves a salty residue. But there are also slower processes at work in taking minerals from the sea, as evidenced by large salt deposits, gypsum, and chemically deposited limestone. In geologic time, sedimentary beds, such as limestone, have been laid down beneath the sea only to rise later, sometimes to form mountains in a process geologists call orogeny.

The importance of salts in the ocean, however, is not where they come from or how they stay there. Since salts are indispensable to life, it is only natural to assume that life began in the oceans and that salts helped the miracle to happen.

Salt for all season. *In its mineral form of halite (left), salt is mined in many nations of the world, while those countries bordering the ocean can obtain salt simply by evaporating seawater.*

Salt dome. *The illustration (below) depicts how salt deposits push up through sedimentary rock layers which are of lower density. Oil deposits are often found in association with salt domes.*

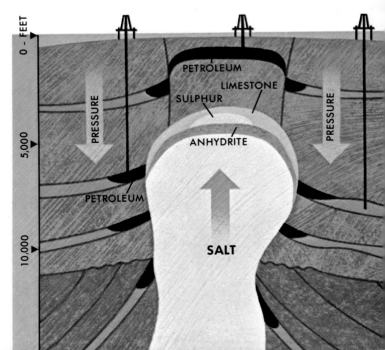

into the rivers, and were eventually carried out to sea. The process continues today, with rain-fed streams leaching the continents of salts and transporting them to the oceans.

It would be wrong to assume, however, that the oceans are just concentrated river water, for there are forces at work on the sea which don't affect rivers. The salinity of the ocean, rather than being continuously built up by the continental runoffs, is kept in a state of balance by various mechanisms withdrawing some of the salts.

The Briny Deep

All the oceans of the world are salty, but some are saltier than others. Tropical waters generally have more salt than temperate or polar seas. This is because the evaporative process powered by the sun's heat extracts pure water, leaving behind a more concentrated solution, especially in enclosed or tropical seas like the Mediterranean or the Red Sea. The average salinity of the oceans is 35 parts per 1000 by weight, but in the Red Sea the salinity may exceed 40 per 1000. In contrast, the polar regions have low salinities because of high precipitation, fusion of ice, and river runoff. One of the least salty seas is the Black Sea, which has a salinity of 18 per 1000 for surface water, increasing to 22 per 1000 in deeper waters.

The importance of salinity is illustrated by its effect on the distribution of life and on the circulation of water in the sea. One classic example of a current controlled by salinity is the flow of deep water into the Atlantic from the Mediterranean. As the surface water in the Mediterranean becomes more saline due to evaporation, it increases in density and sinks. This sinking is balanced by an inflow of surface water from the Atlantic with lower density and lower salinity. To complete the cycle, the deeper, more saline water exits to the Atlantic through the Strait of Gibraltar. Because of the rapid inflow of surface water into the Mediterranean, almost four miles an hour, ancient sailing ships experienced considerable difficulty entering the Atlantic. The Phoenicians discovered that by lowering weighted sails into the deeper outgoing waters, they could easily be carried against the incoming currents and out to the Atlantic.

Isohalines. In polar regions summer surface salinity varies, as measured in parts per thousand.

SALINITY
(0/00)

37+

36-37

35-36

34-35

33-34

33 −

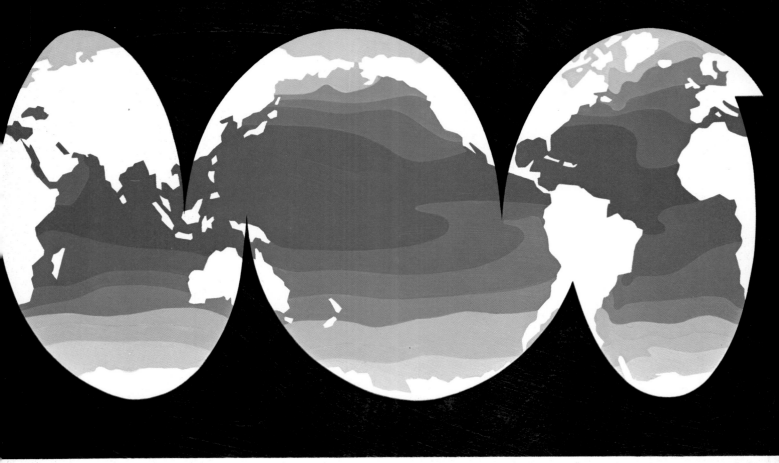

A Matter of Degree

Evaporation, caused by the sun, not only affects salinity, but also cools a thin layer at the surface of the sea. But the sun's radiations are absorbed, especially the red and infrared parts of the spectrum, which in turn heat the water masses. In the tropics, the balance is on the warm side. Ocean temperature varies roughly between 85° in the tropics to below freezing in higher latitudes. The salt content lowers the freezing point, so seawater is still liquid a few degrees below 32°, the point at which fresh water freezes.

Water temperature varies in response to the summer or winter sun, which affects the distribution of life in the sea and is of importance to fishermen. Some tuna seldom venture into the colder water lying below the warm surface waters. If the thermocline—the line of temperature discontinuity be-

Isotherms. Surface temperatures in August are still highest in the tropics, lowest in polar regions.

TEMP. °F.

86+
77-86
68-77
59-68
50-59
41-50
32-41

tween surface and underlying waters—is close enough to the surface, tuna fishermen can expect a greater catch when they set their purse seine nets. After the fishermen encircle the traveling school in their nets, the tuna remain trapped along the perimeter and by the barrier of cold water below, which the fish will not penetrate, even to escape.

Surface waters stay relatively warm to a depth of 300 to 1700 feet. From there, it cools fairly rapidly, in a transitional layer, from 1700 to 3400 feet; further below, water cools very slowly to an almost uniform 32° near the bottom throughout the year and all over the world.

Temperature and salinity combine to determine the density of seawater, which in turn governs the layering of life in the sea.

15

The Universal Solvent

One of water's most important properties is its ability to dissolve gases and minerals necessary for life. Water, in fact, is called the universal solvent because it can dissolve more kinds of material than any other liquid. Two of the most important gases it absorbs are oxygen and carbon dioxide. The carbon dioxide is essential for plant life, which is the beginning of the food chain; oxygen is used for respiration by aquatic animals, and more

Steam bath. Water's gaseous state is induced by this flow of molten lava. The steam comes not only from the ocean but also from the volcanic water.

dissolved oxygen is found in the colder than in the warmer waters.

Even though well-aerated water holds only about one-twentieth the oxygen that an equal amount of air holds, it is sufficient for all cold-blooded ocean animals. Nevertheless, warm-blooded creatures, like sea mammals, that have a high thermodynamic output, need a lot more oxygen than the ocean can supply them and they breathe air.

The oxygen in water is not needed in great amounts for plants during darkness, when they respire like animals. The oxygen is also used to oxidize some organic matter and the rest is carried deeper by currents.

Water can absorb great quantities of carbon dioxide, and this gas is often present in greater concentrations in the sea than it is in air. This ability to handle carbon dioxide is so great, in fact, that the ocean can absorb the gas in polluted areas, such as the North Atlantic, transport it, and release it in less polluted areas, such as the South Atlantic.

Carbon dioxide is used in the production of tissues of both plants and animals and is involved in the development of skeletal matter through a number of chemical compounds, such as calcium carbonate and magnesium carbonate. These carbonates play an important buffer role in the distribution of carbon in the seas.

One other property of water which is important to support life is its transparency. The clearer the water, the deeper in the sea the sunlight penetrates and the thicker the productive layer in which photosynthesis is possible. Accordingly, the global productivity of living matter is affected by turbidity.

The interplay of water's physical and chemical properties combine to keep the sea in a constant state of motion. The water moves as a result of the shifting and settling of different layers of different densities as determined by temperature and salinity. And much of the biological life in the ocean is constantly on the move as it follows the food up, down, and around the world.

Vulcan's contribution. A diver breathing compressed air watches volcano spew. Many of the exotic gases in the sea are of volcanic origin.

Liquid Provinces

As currents silently course through and across the seas, they carry with them great volumes of water, often very different in physical and chemical character and inhabited by specific communities of living creatures. Some currents may bring nutrients to fertilize barren surface waters; some carry schools of fish across great expanses; some transport water thousands of miles to temper harsh climates; and others may collide, piling up great populations of marine life which will not venture into a hostile water mass. In spite of this liquid turmoil, kneaded by currents, there are definable zones in the sea that can be differentiated by their compositions and associated life.

The largest in surface area is the pelagic zone which consists of the upper, brightly sunlit regions of the open sea. This zone is poor in vital nitrates and phosphates. Away from the continents the pelagic zone may be considered as a blue desert where life is sparse and inhabitants wander, or are carried, along the surface. Here and there currents bring

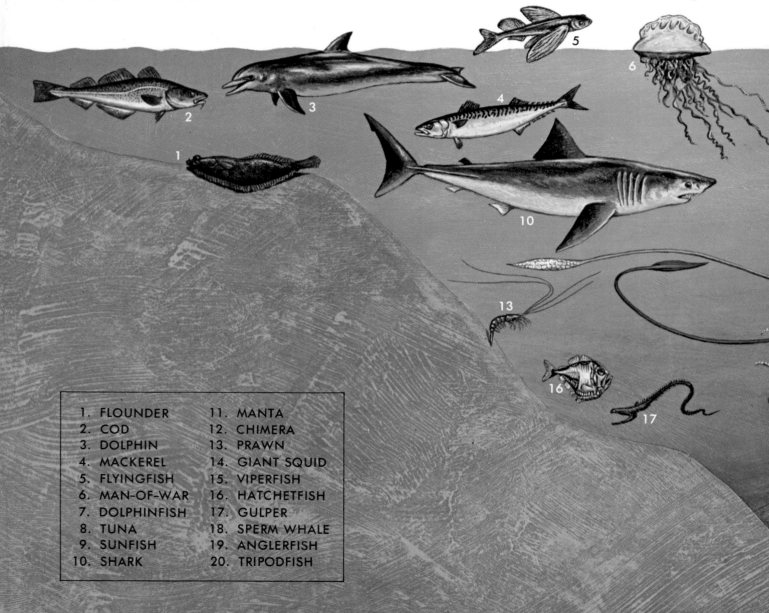

1. FLOUNDER
2. COD
3. DOLPHIN
4. MACKEREL
5. FLYINGFISH
6. MAN-OF-WAR
7. DOLPHINFISH
8. TUNA
9. SUNFISH
10. SHARK
11. MANTA
12. CHIMERA
13. PRAWN
14. GIANT SQUID
15. VIPERFISH
16. HATCHETFISH
17. GULPER
18. SPERM WHALE
19. ANGLERFISH
20. TRIPODFISH

deeper, nutrient-laden water to the surface where phytoplankton utilize sunlight to produce food upon which other life depends. These open-sea pastures can be recognized by their green color and are sought by fishermen as signs of good fishing.

Nearer the continents, in lesser depth, this open-ocean environment is termed the neritic zone and is generally much richer in both abundance and diversity of marine life. It is here where kelp beds and coral reefs flourish, providing protection and food for a host of animals constituting great food webs.

Inseparable from this neritic zone are the littoral and sublittoral zones which include the animals that live attached to the bottom in the tidal region and in deeper waters below the neritic region. Many of the free-swimming animals of the neritic zone depend on the littoral area for protection, food, and nursery grounds. The decomposition and return of organic materials take place in this benthic environment and contribute to the nutrition of the waters above, where the cycle begins all over again.

In deeper waters below the sunlit surface and beyond the continental margins lies the deep sea. It is in these black depths that bizarre bathypelagic fish live above the bottom and weird abyssal organisms sift through the bottom ooze that has rained down from above. All life in this region depends on food falling from above, since photosynthesis cannot take place without light. The bottom-dwellers digest and redigest the sediments, whereupon the bacteria ultimately break down the organic materials and release them as nutrients to enrich the waters. Life has been found at the bottom of the trenches in the deepest seas, in the most hostile environment where there are tons of pressure per square inch of space, the light of day is never seen, the temperatures approach freezing, and food is so scarce that few animals grow beyond a foot in length.

The existence of life in these depths had been questioned until the *Trieste* descended to 35,800 feet and saw fish and shrimp in pressures of 200,000 pounds per square inch.

PHOTIC ZONE

APHOTIC ZONE

Layering of life. There are very few animals that range far and wide throughout the ocean. Out of necessity most have adopted a way of life which restricts them from doing so. For example, some animal live close to shore, others deep in the abyss or in the shallow waters in the middle of the sea. In general, life is most abundant in the areas over the continental shelf and least plentiful in the abyss.

Chapter II. Water in Motion

Standing on the seashore, the wind rustling your hair, the whitecaps foaming over the rocks, it is easy to realize that there are two fluids that play major parts in our lives—air and water. These two fluids may seem very distinct, and to be sure, they do possess different properties, but their similarities and interactions are amazing.

For example, these fluids both circulate around the earth, and their flow is caused by their common property of rising when heated and sinking when cooled. This simple circulation system is broken up by the rotation of the earth and by the influence of landmasses. As a result, there are circular belts, or gyres, of both air and water flow. As these flows turn back on themselves, with cool fluid meeting warm fluid, high-density fluid intercepting low-density fluid, turbu-

---

"Ocean waters slowly circulate around the globe and eventually every water particle traverses the earth's surface."

---

lence is established. Whirlpools, eddies, hurricanes, and typhoons are all fed by this circular movement of fluids. It is the interaction between wind and water, the atmosphere and the hydrosphere, that produces the earth's weather, which determines where rain will fall, where plants will bloom, where deserts will spread.

But however much they might be disturbed or interrupted, the waters of the ocean circulate all around the globe. The movement is slow, but eventually every particle in the water traverses the earth's surface. You have probably seen ink or dye diffuse in a container of water, and with a little stirring it becomes equally distributed everywhere.

The same is true of the oceans. A bottle of ink or a box of salt dropped into the ocean will sooner or later be disseminated equally throughout the sea. The diffusion may take thousands of years. The friction of the winds dragging across the surface waters and action of the eddies and whirlpools are necessary for the stirring. It may take ages, but the substance will be distributed.

In this sense the oceans are great equalizers, for through their motion they are constantly diffusing. They are making the saltiest waters less salty, the warmest waters cooler. Conversely they are constantly bringing salt to the less salty regions and transferring heat to the colder waters. The only reason ocean water isn't perfectly uniform throughout the world is because the sources of fresh water and of heat are beyond the reach of the sea, upsetting the move toward balance.

The constant flow of the oceans always tempted man into finding out where the flow went, where it started. Man utilized currents before he knew what they were. Thor Heyerdahl has shown that it is possible to float halfway across the Pacific in a raft or to sail from Africa to America on a bundle of reeds, using only current and wind as propelling forces. These currents have always been there. But the details of why they are there, how they move, their beginnings and ends are still being determined. We are still perfecting our knowledge of why exactly the water moves, but since antiquity we were able to utilize water in motion.

Tongue of the ocean. The lighter blue water indicates the shallow Bahama Banks which abruptly meet the darker hues of deep ocean water in the Atlantic. These waters become part of the unending circulatory pattern of the sea that occurs thousands of feet deep and at the surface.

Polar circulation. As seawater moves toward the poles, it gets heavier because its temperature drops and density increases. As a result, the body of water moves down toward the bottom.

The Plows of the Ocean

Marine animals living beneath the interface would die if the oxygen they breathe was not somehow replenished. The oxygen-enriching process takes place at the surface interface where the pressure of dissolved gases tends to equal their partial pressure in the atmosphere. The dissolved oxygen is carried throughout the oceans when the oxygen-laden colder waters of the polar regions, especially Antarctica, sink and slowly move across the bottom of the basins.

The movement of deep-ocean water, which is primarily determined by temperature and salinity, is called thermohaline circulation.

The principles involved are very simple. As water is heated, it expands and rises. The colder water is pushed away from the warmer water and sinks because it is denser. Increased salinity also increases the density, and this too helps the water sink in the higher latitudes where ice forms; ice is primarily fresh water because the salts are ejected when seawater crystalizes, which increases the salinity of ambient masses.

In the antarctic region, the dense, cold water sinks and flows slowly northward. The motion is so predictable and the characteristics of the water so distinguishable that the large masses of cold polar water have been given names, such as Antarctic Bottom Water,

Antarctic Intermediate Water, and for the mass that forms in the Norwegian Sea region, North Atlantic Deep Water.

The Antarctic Bottom Water, one of the world's densest bodies of water, moves northward very slowly along the bottom, while the Antarctic Intermediate Water, which is less dense, also moves northward but at a depth of about half a mile to a mile. The smooth flow of these bodies of water is disrupted by underwater mountains and ridges, especially in the mid-Atlantic region.

In the Atlantic, the Northern Deep Water has been traced as far south as 10° below the Equator, where it comes in conflict with the Antarctic Bottom Water, while the Ant-

arctic Intermediate Water has been followed as far north as the West Indies. In all other areas, it is the Antarctic Bottom Water that is dominating, reaching as far north as the northeastern Pacific; the flow of deep water **from the Arctic Ocean** is restricted by the continents, land formations such as islands, and submarine topography.

This thermohaline circulation is an extremely slow process. Radioactivity measurements in this bottom water prove that some of it may be as much as 400 years old in the Atlantic and 1500 years old in the Pacific. But however slow this circulation is, eventually the water on the bottom will again be brought to the surface and into contact with the atmosphere.

Block of sea. Some of the layers of water in the antarctic are seen in this schematic illustration. The differences in density, which determine layering and currents, are due to temperature and salinity.

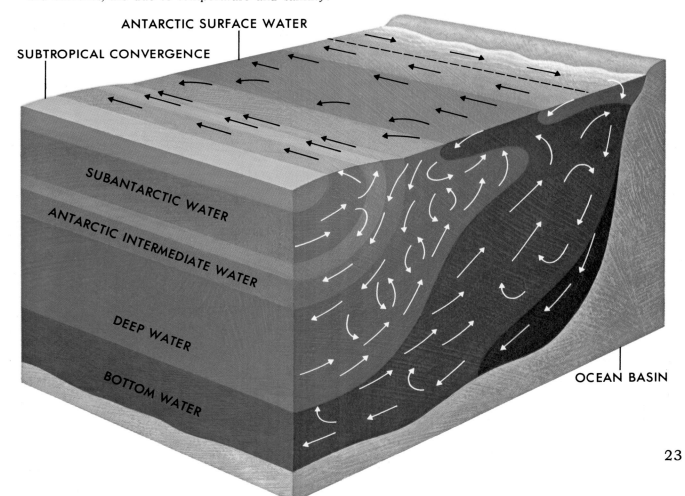

ANTARCTIC SURFACE WATER

SUBTROPICAL CONVERGENCE

SUBANTARCTIC WATER

ANTARCTIC INTERMEDIATE WATER

DEEP WATER

BOTTOM WATER

OCEAN BASIN

Circles of Fertility

When strong offshore winds blow warm surface waters out to sea, cooler subsurface waters then move up and into the shore areas. This is one of the mechanisms that causes "upwelling" and is important because the cooler water brings nutrients from the deep to the surface where they can be utilized in photosynthesis.

Ordinarily there is a layering, or vertical distribution, of nutrients in the ocean. The surface layer, which extends down about 700 feet, is usually low in nutrients because the supply has been depleted by plants; next comes a layer where the amount of nutrients increases greatly, partially because light cannot penetrate and photosynthesis does not take place. The zone of minimum oxygen content is often close to a depth of 2400 feet. Below this is the layer of maximum concentration and on the bottom is a layer of less concentration but more uniformity in the amount of nutrients.

Upwelling disturbs this layering and provides nutrients that are necessary to begin the life cycle. It happens in coastal and equatorial regions on the western margins of continents, off the coasts of California and Peru in the Pacific Ocean or off the coast of Mauritania in the Atlantic.

The distribution of nutrients is not the same everywhere. There are seasonal as well as geographical differences, affected by sunlight and the growth of phytoplankton. Antarctic waters are especially rich in nutrients as the result of strong circulation patterns, which are constantly bringing nutrients into the surface layer. Despite high productivity of plankton, the enormous supply of nutrients in antarctic waters is never and nowhere completely used up.

Commercial fishing. Good fishing is the result of nutrient-rich waters rising to the surface.

Wind Circles

An ill wind may blow no good, but it certainly has an effect on the ocean. It is hard to identify cause and effect in the interaction between the atmosphere and the oceans. The wind, for example, affects the circulation of seawater by generating waves, mixing surface water, and cooling the surface through evaporation. The sea, in turn, warms, cools, or humidifies the air and causes it to move, which will affect the sea locally or in some distant area.

Air circulates just as water does: warm air rises at the equator because it is heated and it expands. This produces a low pressure area, and cooler, denser air flows toward the area. Air, like water, is cooled, contracts, and sinks in the higher latitudes.

Both air and water circulations are complicated by the rotation of the earth. Instead of a smooth, steady flow, the fluids are deflected to the right in the Northern Hemisphere and to the left in the Southern Hemisphere. This is known as the Coriolis effect, named for the nineteenth-century French mathematician who discovered it.

Under the influence of the Coriolis forces, the air, instead of simply rising and flowing

directly north, then descending and flowing south, tends to circulate in an elliptical manner, oriented at an angle off north and south. For example, the air that ascended at the equator in the Northern Hemisphere veers to the right and ends up descending east of where it was heated. The return of this cool air to the equator is also deflected to the right. The result is winds that blow from the north and veer to the west, which we call the northeast trade winds. A similar system exists at the poles, where air is heated, expands and rises, then cools and becomes heavier and descends, all the while being deflected as the earth turns. Between these tropical and polar cells of air moving elliptically, there exists another cell of air that behaves differently from the others. It acts like a gear and travels in a direction opposite to that of the others. It is pulled down with the descending tropical air, but then it travels northward to rise with the ascending air of the polar cell and completes its cycle. Of course, it is deflected by the Coriolis force and gives rise to what we call the prevailing westerlies. Wherever these cells of moving air come into contact, there is an exchange of heat from the warmer to the colder one.

This atmospheric circulation is felt on the earth's surface in the form of winds, which are called easterlies at the poles, westerlies in the temperate zones, and trade winds (also easterlies) in the tropics. Since air is rising at the equator, there is very little surface wind. This belt of windless ocean is called the doldrums, named by the sailors of old who were often becalmed there and waited for days for a breeze to power their ships back to the steady trade winds.

Interaction. The sea (top, left) affects winds by heating, cooling, and humidifying air.

No power. A becalmed sailing ship (opposite page) may have to wait days for a wind to come up.

GREENLAND

SPANISH
GALLEONS
(C)

EUROPE

NORTH AMERICA

MANILA
GALLEONS
(B)

EQUATOR

LA BALSA
(A)

SOUTH
AMERICA

AFRICA

RA II
(D)

6
7
5
4
3
2
1
13
17
12
11
14
15
16
10
8
9
19
18

A.
B.
C.
D.

Water Circles

Since man first began sailing the seas, he has been aware of and frightened by the surface currents of the oceans. These streams within the ocean carried him across the waters and home again, usually in a circular path rather than in a straight line.

In 1812 Alexander von Humboldt detailed the evidence for a flow of cold water below warm tropical seas. He said this showed that, in addition to the general, worldwide circulation of water, there were streams which were distinct from the seas through which they flowed. The currents are distinguished by such properties as velocity and direction, and they have distinct boundaries. Usually currents are linear in form, but they may also be broad or sheetlike.

Most of the oceanic currents are caused either by wind friction on the surface of the sea or by gravity acting on water of different

28

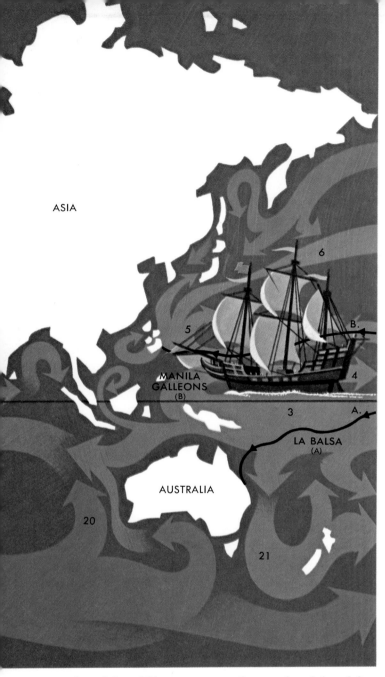

ASIA

6

5

B.

MANILA
GALLEONS
(B)

4

3

A.

LA BALSA
(A)

AUSTRALIA

20

21

eastern margins of the continents, they become among the strongest and largest currents in the world. The Gulf Stream off the Atlantic Coast of the United States and the Kuroshio, or Japanese, Current off the Asian landmass are equatorial currents.

The direction of surface currents, besides being affected by wind and landmasses, is also altered by bottom topography and by the Coriolis effect, which deflects water to the right in the Northern Hemisphere and to the left in the Southern Hemisphere.

The surface currents flow slowly: at an average of less than two miles an hour, in some regions the waters seem to stand still. But in a powerful current like the Gulf Stream, the average velocity is three miles an hour and parts of it have been measured at six miles an hour. The width of the Gulf Stream varies from 50 to 150 miles, its depth from 1500 to 5000 feet, and it has a flow of 70 million tons of water per second.

Natural power. *Early navigators had to depend on surface currents and associated winds in order to cross the ocean. Some crafts were primitively simple, such as La Balsa (A), which went from Peru to Australia, or Ra II (D), which sailed from Africa to Barbados. More sophisticated were the Spanish galleons (C), which carried on trade between Europe and America, and Manila galleons (B), the Pacific between Mexico and the Philippines.*

densities. The currents determined by friction, or wind drag, generally form loops or gyres matching the wind patterns. In the Atlantic and Pacific oceans these currents are deflected toward the poles after bumping into the continental landmasses. The trade winds, which blow from east to west, form the equatorial currents found in all the oceans. When these equatorial currents are deflected by land, they are accelerated as if in an injector nozzle. As they flow along the

WARM CURRENTS	COOL CURRENTS
1 ANTARCTIC WEST WIND DRIFT	12. NORTH EQUATORIAL CURRENT
2. PERU CURRENT (HUMBOLDT)	13. GULF STREAM
3. SOUTH EQUATORIAL CURRENT	14. NORWEGIAN CURRENT
4. EQUATORIAL COUNTER CURRENT	15. NORTH ATLANTIC CURRENT
5. NORTH EQUATORIAL CURRENT	16. CANARIES CURRENT
6. KUROSHIO	17. SARGASSO SEA
7. CALIFORNIA CURRENT	18. MONSOON DRIFT
8. BRAZIL CURRENT	(SUMMER EAST, WINTER WEST)
9. BENGUELA CURRENT	19. MOZAMBIQUE CURRENT
10. SOUTH EQUATORIAL CURRENT	20. WEST AUSTRALIAN CURRENT
11. GUINEA CURRENT	21. EAST AUSTRALIAN CURRENT

Chapter III. Border States

A fetching maiden. This term often describes a demure young woman. But fetch, referring to the sea, is anything but demure. Fetch is the amount of open water across which a wind blows. For the wind is the source, origin, genesis of most of the ocean's waves whether they are the multidirected ripples formed by gentle breezes, the real waves formed by ten-mile-an-hour winds, or the savagely high seas piled up by gales blowing over large stretches. The size of the waves raised by wind depends on its velocity, on the amount of time it blows, and on the extent of the surface it acts on. These are the factors which start the sea rolling, a movement that can reach to the very edge of the ocean basins.

A wave being pushed toward the shore demonstrates the theories of inertia and motion. For as the rolling, circular motion of the water is continuously transferred without interruption, the amplitude, or height, of the wave is determined by the force of the wind. But as the bottom rises, such as near the shoreline or where there are ridges, the wave

---

"These rip currents, as fast as two miles an hour, can easily trap an unwary swimmer."

---

is tripped up and spills head over heels onto to the shore, like a line of ice skaters who suddenly run into sand and are slowed from below as a result of increased friction, but at the same time are pushed ahead by those who follow, which sends them tumbling.

How big do waves get? Sailors and fishermen have never been known to limit their measurements by the truth, but one of the largest waves ever recorded was reported by the American tanker U.S.S. *Ramapo* in 1933 while on a trip from Manila to San Diego. Lt. Cmdr. R. P. Whitemarsh said the vessel encountered "a disturbance that was not localized like a typhoon ... but permitted an unobstructed fetch of thousands of miles." The wind was estimated at well over 60 knots, and the wave which crashed over the *Ramapo* hit well above the mainmast crow's nest. Whitemarsh estimated the height of the wave to be 112 feet. Today satellites can accurately measure wave heights as well as open-sea tides, and eventually knowledge will replace tales and legends.

Waves do not always meet a beach head-on. Often they approach at an angle, but there is a tendency to swing around toward the shore, called wave refraction. This angle adjustment toward the beach is not complete, and the result is a longshore current which flows along the beach. This current increases until it can overcome incoming waves, and then it flows back into the sea in a rip current. These rip currents, as fast as two miles an hour, can easily trap an unwary swimmer. It is almost futile to swim against a rip current, and it is better to try to swim across it or even ride it out from shore until it slackens before beginning the swim back to shore.

There are also waves that are measured not so much for their speed, which can reach 45 miles an hour, or their breakers, which can be as high as 60 feet, but as generators of damage. These are the catastrophic tsunamis, usually caused by earthquakes or volcanoes, which have killed thousands of people in a matter of minutes.

Riding the curl. Ocean waves range from the gentle ripples to the powerful shipwreckers fed by the fury of storms. In between are the curling breakers that provide momentum for surfers.

Giving Waves a Break

Waves pound toward the shore, carrying surfers on their exciting way. The waves continue pounding after the surfers have gone, incessantly crashing against the rocks and sand. When ocean waves hit the shore, they represent a tremendous amount of energy borrowed from the wind and accumulated over hours. But this energy becomes most obvious in shallow water.

In the open ocean the first puffs of wind cause little ripples going in almost every direction. As these ripples sort themselves out according to size, the waves get bigger, and they amplify each other as they overtake one another. There is a lot of motion involved, but the water itself is not traveling ahead. Water particles are moving in a circular pattern or a back-and-forth elliptical pattern, but they are not moving forward. What is being passed on is the motion itself, as though it were a chain reaction much like a row of dominoes falling down in succession after one had been pushed.

This theory can be tested in any pond by dropping a stone into it and watching the waves. The water itself isn't traveling, but

the wave motion is passed on until it bounces on the rim. If a cork is placed in the water and another stone is dropped, the cork will bob up and down, back, up, forward, and down, as each wave passes under it.

Ocean waves are much more complex. They are made up of swells, which remain active long after the winds that created them have vanished, so that in high seas, sailors often fight two or more crossing swells. But as the waves near shore, their characteristics are altered, under the influence of the ocean bottom. As the water becomes shallower, the ocean floor interferes with the circular water pattern, slowing down the lower part as it drags across the bottom. Water piles up behind the top of the wave, causing the wave to break, flinging itself toward shore. Or, to put it another way, when the wave can "feel" the bottom, the wavelength and velocity are decreased and the height is increased. The wave breaks when the particle velocity of the crest is greater than the particle velocity of the rest of the wave. When the ratio of the height of a wave to the depth of the water is about three to four, it will break. How the wave breaks depends on the slope of the ocean bottom.

Wave dynamics. *As a wave approaches shore the drag of the bottom shortens the wavelength from A to B to C. The bottom begins to affect the wave when the depth of the water (D) is one-half the wavelength (A). As this shortening of wavelength takes place the wave height consequently increases.*

The waves seem to bunch-up as they approach the shore. Finally when the wave height (E) reaches a ratio of three to four with water depth (F), the wave tumbles over on itself or breaks. Upon reaching a sloping bottom, waves spill forward; a steep bottom causes them to leap into a plunging wave.

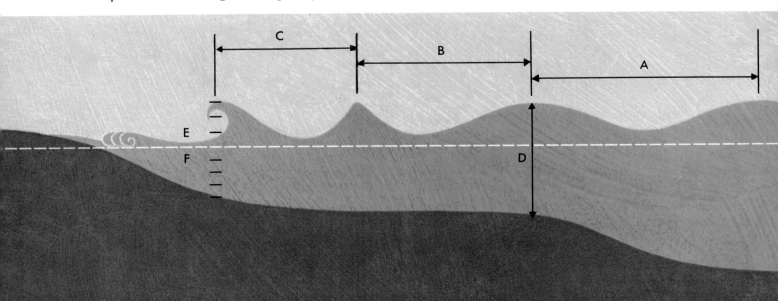

The Wavy Seas

Breaking waves may mean fun to a surfer or a beautiful picture to a photographer, but they can mean danger to a boatman for they signify that he is approaching land. Often, along low coastlines or on islands surrounded by bars and reefs, breaking waves are the first indication that land is near.

Breaking waves are called surf, as distinguished from the other types of wind-driven waves—sea and swell.

Sea waves are directly affected by the wind in open water, and swell is the term applied to waves once they form a regular pattern and start to move away from their point of origin. Swell can cover a great distance. Oceanographers have tracked waves generated by winter storms in the antarctic region all the way through the Pacific Ocean until they have broken on the coast of Alaska.

When swells reach shore they are affected by the bottom resulting in a drag to their forward motion and ultimate toppling over or breaking. During those few moments before the wave crashes a very steep face of water occurs and it is this precise moment that the surfer finds optimum for riding waves.

There is another type of wave which no one has ever seen and which is not generated by the wind. This is the internal wave, a common phenomenon occurring between layers of water of different densities. On a calm day evidence of internal waves may be seen on the surface as long lines or patterns of slicks. These coincide with the wave and are positioned between the trough and crest of succeeding waves.

Storm at sea. The research vessel Eltanin struggles against the onslaught of wind and swell while studying the Antarctic seas.

Catastrophe at Sea

When the ocean is at its most spectacular, it is often at its most dangerous. The stark majesty of antarctic waters, the beautiful but poisonous fish of tropical waters, or the awesome and destructive tsunamis are all threats for man. The tsunamis are often mislabeled tidal waves, but they have nothing to do with tides since they are caused by volcanic eruptions or some movement of the ocean floor, such as an earthquake.

Tsunamis can have extremely long wavelengths, of over 100 miles, and travel at speeds over 350 knots, so when they begin to drag on the bottom in shallow water and break on the shore, they can achieve incredible height and destructiveness. Like all breaking waves, tsunamis are affected by the topography of the bottom, breaking highest

near submarine ridges and breaking the least near points of land surrounded by deep water. Since the force of a breaking wave can be controlled by the sea floor, in some areas the full force of a tsunami might be channeled and concentrated into a small area, while in other places it might be dispersed.

Another type of catastrophic wave is the storm surge, which consists of a steady rise in the water level rather than a rapidly rhythmic rise and fall of water. Strong winds, such as those associated with hurricanes and typhoons, can pile up water on the coast. The high-water levels due to these storm surges can be especially damaging if they coincide with high tide in low coastal areas.

The stationary wave can be equally destructive in enclosed bodies of water such as a bay or, more commonly, a freshwater lake. Unlike the fast-moving undulation of a swell, in a stationary wave the surface moves up and down around the edges, much like the movement in a bowl of soup that is tilted and then placed level again. Storms and rapid changes in atmospheric conditions are the usual causes of stationary waves.

A fourth type of natural catastrophic wave is the landslide surge, caused by a large amount of rocks or ice ripped loose by landslides or glaciers and plunging into the ocean.

Huge surf batters the north shore of Oahu (opposite, top). Such fury is periodically unleashed on this unprotected coast causing extensive damage.

A Chilean earthquake generated a tsunami which caused damage in Hilo, Hawaii (opposite, bottom), and as far away as Japan and Australia.

Cyclone and waves. Boats (below) were beached by tsunamilike waves which were whipped up by 80-mile-an-hour winds on the Bay of Bengal.

Chapter IV. Ins and Outs of Tides

Tides are waves that occur every 12 hours and 25 minutes and that have a wavelength of half the circumference of the earth, approximately 12,600 miles.

But this definition isn't complete. It does not explain why, for example, there are two tides a day (semidiurnal tides) in some areas and only one tide a day (diurnal tides) in others. Nor does it explain why the difference between high tide and low tide can be less than two feet in the Gulf of Mexico and more than 50 feet on parts of Canada's Atlantic coast. As the tides sweep around the globe, water is actually moved sideways, and the resulting water levels are usually different on opposite sides of a continent.

"The ancient Greeks realized there was a connection between the moon and tides even though they could not absolutely determine the cause and effect."

Many of the early explanations for tides were simplistic—such as tides representing the breathing or pulse of the ocean. But the ancient Greeks realized there was a connection between the moon and tides even if they could not determine the cause and effect. Although well outlined by the poet John Donne as early as 1618, a rational explanation had to wait until Sir Isaac Newton formulated his theories about gravity. Now we know that the moon attracts the waters of the oceans and that the sun has an effect too, but because of its much greater distance from the earth, its gravitational pull is much less than the moon's. Sometimes the sun's force reinforces the moon's pull, and sometimes they work against each other, alternately causing higher tides and lower tides.

But even this doesn't fully explain the form tides take and their effect on living organisms. There are also such factors as the rotation of the earth, bottom topography, shoreline configuration, and depth of water.

The study of tides is aided by the wizardry of modern computers, but practical tidal predictions are still being done by the age-old method of daily observation, year in and year out. Such observations have enabled ports to be built in areas where there are 20-foot tides, or allowed fishermen to determine when and where certain fish can best be caught, or shown how to use the ebb and flow for maintaining ship and ferry schedules.

The constant coming and going of tides can establish currents which may bring deep water up to the surface. In coastal seas or narrow passages such as those in the Aleutian Islands off Alaska, tidal currents can be swift and strong enough to endanger ships. Tidal currents are relatively weak in the open sea, but in the shore areas they may be strong enough to scour harbors by removing large amounts of sediment and thus eliminate the need for dredging. But this sediment might well be deposited in some other area, blocking a different channel or harbor.

The complex nature of tides is more than a scientific curiosity, for they directly affect the lives of everyone from harbor masters to transoceanic travelers, from fishermen to beachfront property owners. Indirectly, the daily tides affect the lives of everyone living in a coastal region.

Tidal lifestyles. The certainty of tides is such that ocean organisms have adopted habits based on their ebb and flow. These mussels and barnacles actively feed when the water is at high tide, but they close up tightly to retain moisture when the tide is out.

Ups and Downs of Tides

Simply speaking, tides are caused by the gravitational pull of the moon, and to a lesser extent, the sun, on the surface of the earth. As the earth rotates, the point of the ocean closest to the moon is more strongly attracted than water farther away, and the result is a "bulge" in the ocean. At the same time, on the other side of the globe, the centrifugal force caused by the earth's rotation causes the water to bulge at a point farthest from the moon. High tides are thus generated on opposite sides of the earth.

As the earth continues to rotate, water recedes from our original bulge and moves continuously to swell other parts of the ocean. When the sun is aligned with the moon, at times of a new moon and of a full moon, the gravitational forces of these bodies act in concert to produce the highest tides, called spring tides, which can occur in seasons other than spring. When the moon and sun are at right angles to each other, when the moon is in its first and third quarters, the gravitational pulls of the two bodies work against each other and the tides are lower than usual. These are called neap tides.

Another factor affecting the height of tides is the elliptical orbit of the moon, which at times brings it closer to the earth. Twice a year the moon passes very close to the earth causing very high perigee spring tides, and twice a year, when the moon is farthest away, there are apogee tides of minimal height.

The influence of the moon on tides being much greater than that of the sun, the time cycle of tides is 24 hours and 50 minutes, a lunar revolution, rather than the 24 hours of a solar revolution.

Mont-St.-Michel. This tiny French community can be approached on land only at times of low tide.

Special Effects

For want of a nail a kingdom was lost. That's what they said when the king's horse lost a shoe, and the king was unable to lead his troops into battle. For want of a tide table, many another battle has been lost. Julius Caesar lost his boats on the shores of England, when he was unaware how high the tides were and his men failed to pull their boats far enough ashore to safety. But after all, what did Caesar know about tides? In the Mediterranean and Adriatic Seas, the tides don't vary much more than a foot or two, and few of his sailors knew that in other places tides were greater.

When tide waves from different points move through the ocean, they can conflict with each other, produce interferences, and lessen the rise and fall of tides along some coasts. In other instances, the waves join force, come in resonance, amplify each other, and precipitate great differences in high and low tides. Shoreline shape and bottom configuration can cause great ranges in tide height.

In restricted areas, such as bays and channels, there can be large and rapid buildups of water. This occurs as the tide waves approach the shore, they encounter narrowing passages that restrict entry and cause the water to pile up. This results in a tidal current, which precipitates a very much higher and more rapid tidal rise.

One of the most spectacular displays of this is in the Bay of Fundy in Canada, between New Brunswick and Nova Scotia, where there is a 50-foot rise of water in about 10 minutes.

In circumstances where tidal movements are restricted before reaching the shore, they may even rush up coastal streams. These tide waves, called tidal bores, move upstream for several miles, sometimes at a speed faster than 10 miles an hour. The city of Cologne in Germany, located 120 miles inland, is a seaport because oceangoing ships can navigate up the Rhine River when the tide is flowing upstream and inland.

Bay of Fundy. *Extreme tidal rises in the Bay of Fundy can be seen in Port Williams, Nova Scotia.*

Action and Reaction

The rhythm of life has often been related to extraterrestial events—like the cycle of the moon. Some thinkers saw a connection between the menstrual cycle of women and the lunar cycle. Charles Darwin took the connection of the moon, tides, and menstrual cycles to show that life had indeed originated in the sea and that women were still showing the effects of the tidal cycle common to many ocean organisms.

There are creatures (some shellfish, for example) that ordinarily live in tidal areas which, when removed from the ocean and put into a nontidal environment like an aquarium, continue their tidal life-styles.

In the open ocean, tidal effects are of little or no consequence to pelagic organisms, but those animals which live near the shore must do daily combat with the tides. Mussels and barnacles, for example, simply close up tightly and wait for the water to return. Certain gastropods are able to secrete a mucous coating that effectively prevents desiccation, or drying out. Other animals burrow into the muddy or sandy bottom to await high tide, while some organisms remain in the shallow tide pools that are left by the outgoing sea.

A more bizarre response to tides is displayed by *Convoluta,* a small flatworm which lies exposed on the sandy bottom during low tides at daylight hours. The flatworm uses this opportunity to bare itself to the sun so that the greenish algae that live symbiotically in its digestive tract may use the light to grow. But as soon as the tide starts to return, *Convoluta* returns to cover in the sand.

Not all tidal responses are daily. Some organisms' behavior is attuned to longer tidal periods, such as the spring and neap tide cycle. One of the best known of these is the grunion's spawning activity. This fish deposits its eggs on the sandy beach near the high water mark during the spring tide. The next time the water approaches that mark, during the next spring tide, the grunion eggs are ready to hatch and the larvae are carried out to sea with the receding tide.

The earth itself reacts to tides, since as the water moves to and fro there is friction against the rotating globe. This causes a slowdown in the speed of revolution of about 1/1000th of a second every hundred years. There is also a corresponding speed-up in the moon's rotation which results in the moon moving away from the earth and the length of the days becoming slightly longer.

Spawning grunion. On the three or four nights following a full moon, the beaches of southern California are littered with the long, silvery bodies of grunion which have come ashore to breed. This phenomenon occurs only from March to September, when the spring tides are receding from their highest.

Chapter V. Threshold of the Sea

Land and sea appear to be in almost constant conflict, each seeking to overtake the other's domain. The pounding sea eats away at a coastline along America's Pacific Coast, while the Nile River is busily depositing silt in the Mediterranean Sea, extending Africa's boundary ever northward, a bit at a time. Rather than being in conflict, land and sea are in a state of dynamic equilibrium, for the total sum of land on the earth changes little, however dramatic the gain or loss might be in any specific place.

Man has complicated matters, for he seems to be a creature that prefers permanent boundaries and static conditions vis-à-vis the sea. But in his efforts to stabilize shifting coastlines, he has often unwittingly contributed toward hastening erosion. These efforts may be offset by reclamation efforts, such as

---

"Continents float more or less according to the composition and weight of underlying rocks."

---

the spectacular dike system of the Dutch in holding back the North Sea. Half the people of Holland, one of the most densely populated countries in the world, live and work below sea level because of projects like the one that isolated the Zuider Zee, formerly an arm of the North Sea. But the reclamation of land is not without its price, for by prohibiting freshwater rivers from free access to the ocean, the Dutch face the possibility of their impounded water becoming highly polluted. And the dikes prevent tides from reaching inland areas, thus eliminating the periodic flush of salt and fresh water needed by such commercially profitable shellfish as mussels, oysters, and lobsters.

The battle for territory between land and sea is affected by a geophysical process obeying the laws of isostasy, which helps keep large amounts of land above sea level. Continents float more or less according to the composition and weight of underlying rocks. As a result they are eroded by winds, waves, and raging rivers. These rivers carry tons of sediment down to the ocean, where it may be deposited in delta silt or settle out on the bottom. Once on the floor, wave action could bring it back up onto a beach or bottom currents could move it farther out to sea.

An important part of the land particles carried to the sea are carbonates, especially calcium carbonate, which many marine organisms use to form their shells and skeletons. Much of this calcium becomes part of the "calcium budget" of the sea, which is as necessary as the "carbon budget" for continued life. As organisms die and their shells and skeletons begin to sink, some of the calcium is redissolved to be recycled and used again, while the rest settles on the bottom to await processing which turns it into such sedimentary rock as dolomite. Eventually this rock is thrust up again above sea level. It is then subject to weathering and wave action and is again washed into the ocean.

There is much in the sea that is cyclical, whether it is the sculpting of shorelines and rebuilding of beaches, the periodic ebb and flow of the tides, or the pattern of life which depends on change, the coming and going of seawater, fresh water, and atmospheric exposure. The result is a seashore environment that is never exactly the same two days, two years, or two centuries in a row.

Give-and-take. Wave action cuts into shorelines and rivers eat at the heartland of continents, but all is not lost, for there are coastal areas where sediment is trapped and land is built up from the sea. (Pt. Reyes, California).

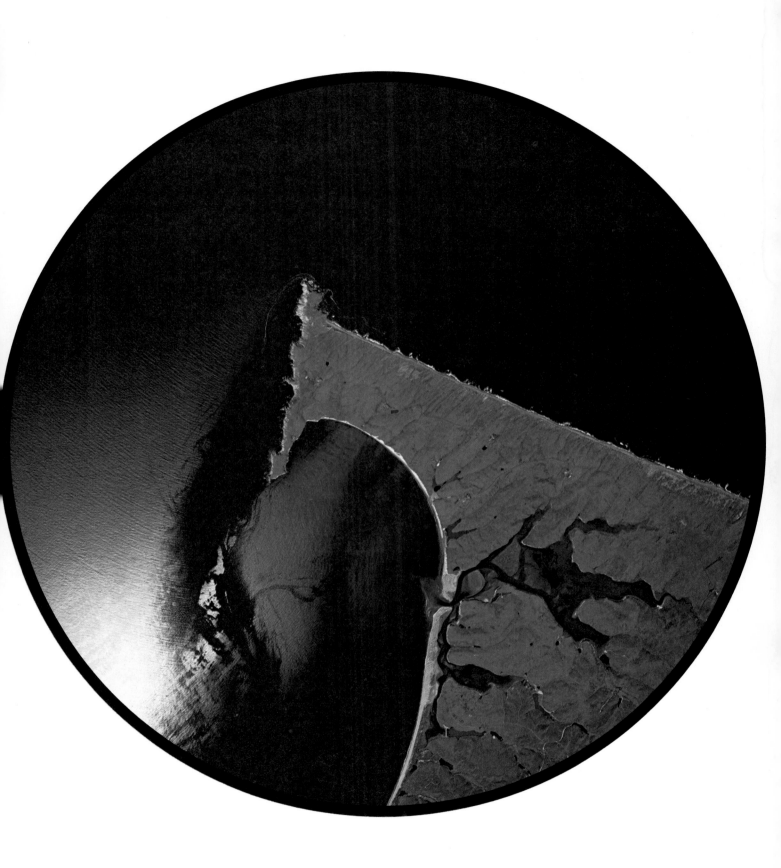

New York. *The sandy beaches of Fire Island in the Atlantic are constantly undergoing reshaping by eroding waves and the deposition of sand.*

Putting on a New Face

The profile of the shoreline is constantly changing; sometimes the alterations are almost imperceptible to the naked eye, and at other times features disappear overnight in a hurricane or heavy sea.

Much of the Atlantic Coast of the United States, from Long Island to Florida and through the Gulf of Mexico, is characterized by barrier beaches—constantly shifting sand islands, which are able to survive because of their malleability. Such islands are found only along continental coasts of sand and gravel, which have gentle slopes. In contrast, the Pacific shoreline of America is rela-tively unprotected and rocky, which provides different features, different problems.

These differences in shore topography are a result of the processes that uplift the continents and shift them over the face of the earth. The Atlantic shoreline is very old and largely made up of material eroded from the land, for example, the Appalachian Mountains. The Pacific shoreline is much more active and is a very recent feature, geologically speaking. Some current theories hold that this is the result of the Pacific Ocean floor, or Pacific plate, being pushed against the North American landmass. Where these two plates meet, wrinkles are produced in the crust, forming a mountainous shore.

California. The rocky Pacific coast is subjected to much more furious pounding by waves than is a gently sloping shore with soft, sandy beaches.

An important feature of East Coast beaches is the berm, that area from the base of the dunes to the high tide lines, that is formed by waves. This is an unstable feature and may be removed by storm surf or just as easily reformed by less turbulent waves. On Fire Island, a 30-mile barrier island off Long Island, the western tip is being extended at a rate of 212 feet each year, due to the littoral drifting of sand. A lighthouse built on the tip in 1858 is now five miles inland.

Such drifting can also be found on the beaches of the Pacific Coast, but for the most part the West Coast is a series of rocky cliffs ranging in a nearly unbroken line. Because of the steeper slope, the surf pounds the rocky shores much more furiously than it would a soft, sloping sandy beach. Unlike the Atlantic Coast, with its bays and inlets, the Pacific shore is almost a straight line all the way to Puget Sound, where the coast becomes irregular and islands dot the area.

The marine organisms living in two such disparate habitats form very different populations. To survive in the pounding Pacific, creatures must be able to attach themselves firmly to the rocks in order not to be swept away. But on the beaches of the Atlantic, the action of the surf on shifting sand poses different problems: the organisms have few hard surfaces available and must burrow below the surface of the sand.

Give-and-Take

What the ocean erodes or eats away in one place is deposited elsewhere and will probably find its way above water at some other place, perhaps at some later time.

Sandy beaches are generally eroded by the rougher waves of winter months, but this is usually offset by the deposition of sand by the gentler waves of summer. Some underwater feature, such as a submarine canyon, could funnel sand to the deep sea, in which case the beach would gradually be eroded.

On rockier coasts, a common feature of erosion is the wave-cut cliff. These usually occur where the shore slopes steeply and the waves break directly on the shore. The wave action produces a sheer cliff and, as it eats further into the shore, produces a terrace or plaza at the base of the cliff. The further back the cliff moves, the longer it takes to eat away more rock. And the type of rock, its hardness or softness, also helps determine the rate of erosion.

As waves approach the shore at an angle, each wave picks up some of the surface grains of sand and washes them a short distance to the side. The succeeding waves do the same and result in a drift of sand along the beach called littoral drift.

In the process of maintaining the balance between land and water, man has undertaken much artificial reclamation, such as the landfills of Manhattan Island in New York and of Fontvieille in Monaco where the real estate is among the most expensive in the world; draining the Pontine marshes in Italy; or building the networks of dikes which reclaimed farmland for the Dutch.

Dutch reclamation. *Sluice caissons are tugged into place to cut off sea flow at Veerse Gat in the Netherlands.*

Continents Afloat

Crinkle up a piece of paper or wrinkle a sheet of aluminum foil. The surfaces that were once smooth and flat are now covered with valleys and hills of different depths and different elevations. But it is impossible to crinkle water. Water is unable to support a static or stable elevation. Yet water, in the form of ice, is able to rise above the surface of the ocean. Because ice has a lower density than water, icebergs are able to float in the sea with their tips protruding. The bulk of an iceberg is below water, but there is always between 15 and 25 percent of its volume protruding. And if that tip starts to erode, the iceberg slowly rises so that part of it is always above the waterline.

On the continents, we see the mountains ranging above the land, and in some of these mountains we find evidence that they were once at sea level. Seashells and beach sand

have been found, indicating that the land has been raised a considerable distance, more than could be explained by a higher sea level caused by the melting of icecaps in the inter-glaciation periods.

The principle that keeps the tip of the iceberg above the water also works on land-masses—provided the base, or "roots," of the mountains are made of rocks with a very low density. Thus, the landmasses are able

Illustrating a principle. Part of the Calypso (opposite page) is hidden beneath the waterline. This demonstrates Archimedes' principle that a body in water is buoyed up by a force equal to the weight of the water it displaces. A similar principle, when applied to continents, is called isostasy (below) and explains how continents are able to stay above sea level. The land areas of the earth are made up of light-density rocks called sial, which "float" on the much denser basaltic rocks that lie beneath them. Underneath the continents, between the sial and sima, there is an abrupt change in the types of rock; this is called the Conrad Discontinuity. There is a similar change, the Mohorovicic Discontinuity, between the sima and rocks of the earth's mantle.

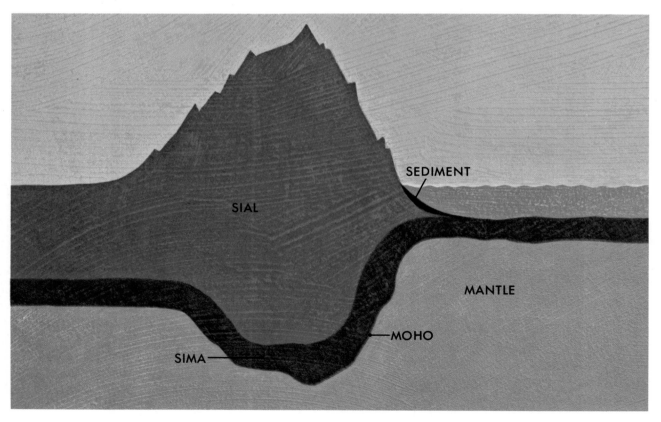

to rebound when part of them is eroded. This is called isostatic rebound.

Continents are not floating about in the ocean like icebergs, for both the oceans and continents are part of the earth's crust, which rests on the mantle, or subsurface layer of the earth. But between the ocean basins and the mantle is a layer of very dense basalt rock, while under the continents there is a layer of much lower density granite rocks. Thus it is the continents that rebound and maintain the isostatic equilibrium. If all the land above sea level were scraped off and thrown into the oceans, this equilibrium would become unbalanced and the continents would rise high enough to set the scales even again. Isostasy helps counteract the effects of erosion, but it does not fully explain why and how the continents have continued to exist since the formation of the earth. This problem will be considered in Chapter XI. But isostasy does help explain why the continents are not eroded flat and covered by shallow seas.

Pulling a plug. *A corer penetrates into the seabed for a sample, or plug, of sediment.*

Settling Down

Rivers, waves, even the air erode the continents and distribute tiny particles of land in the ocean, where many settle on the bottom. The mix includes everything from fertile black topsoil, used to grow foodstuffs in the heartland of a continent, to marine debris, such as bits and pieces of dead organisms. Often the inorganic particles are making a round trip, having been sand at one time, then being deposited, compressed and thrust up as sandstone, only to be worn down again by a cutting river or a biting wave.

These particles are deposited on the shelf of continental land that extends under the ocean. The shelves in the Atlantic Ocean may extend for a 100 miles, while in the Pacific the American continental shelf is very narrow. The sediment on these shelves is sometimes swept back up on the shore, or it might be moved further out to sea for further geological processing. There are many places on the ocean floor where no sediment is found because strong bottom currents remove the particles as fast as they settle out.

When waterborne particles begin to sink, the coarser grains deposit more rapidly than the smooth, fine grains. Turbulence is obviously another factor in the rate of settling, this rate being greater in still water than in active water. We might expect then to find an almost perfectly graded distribution pattern among sediments, with the coarser grains being closest to shore and the finer particles further out. But because of wave energy and sea level changes, this is not always the case; in some areas we find coarser particles far offshore and mud deposits closer in. Features like barrier islands protect the intercoastal waters so that silt and clay are able to deposit. But on the other side of the islands, where the wave action is so much stronger, only the rougher particles settle out.

The picture is complicated by the fact that the sea level rises when glaciers melt and many earlier deposits are buried. Glaciers deposited rocks, pebbles, and particles of all sizes as they retreated. In some cases, these deposits were covered by new sediments carried in by rivers, but in other areas there is little addition of new sediments. Glaciers deposit rocks on the sea floor in another, more indirect way. The moving ice plucks rocks from the mountains and lands through which it travels, and when the glacier reaches the sea, chunks of ice often break off and when they melt, drop the boulders on the bottom. The result is a very fine sedimentary bed with a few large rocks scattered about.

Many sea mammals, like sea lions or fur seals, swallow stones to help their digestion or to prevent hunger during migrations. They may reject such stones in the high seas, and when these stones are dredged, they may puzzle geologists and oceanographers.

Life among the rocks. *Anemones cluster around a rock that was dropped by a melting iceberg.*

Marriage of Waters

The retreat of the glaciers from the last ice age and the subsequent rise of the seas have created some unusual features where bodies of fresh water and salt water meet—the deltas and estuaries.

Rivers like the Nile and Mississippi deposit silt and clay at their mouths, building new land with the river channel in the middle. In time, the river splits into several smaller channels, giving the deltas their characteristic "bird footprint" pattern. Only 3000 years ago the Tigris and Euphrates rivers entered the Persian Gulf as two separate streams, but they added so much land, more than 100 miles, that they now meet inland and empty into the Gulf as a new river.

Some estuaries are simply a glacial river valley whose mouth has been flooded by the rising sea. The upper walls of the valley form a semienclosed body of water and this affords protection for a highly productive biological area. There is not a wide variety of species living in the estuary, but there is a large number of individuals in each existing species. In addition to the permanent residents, the estuary is used as a nursery by many ocean animals that spend most of their lives in open water.

An estuary is a fragile place, one that can exist only in times of rising sea levels and one whose most consistent feature is change. The shallow, sunlit waters are constantly being fed by the nutrient-rich flow of the river and the nourishment of the sea provided by the periodic tides. But with too much fresh water, such as occurs when spring rains are especially heavy, or too much seawater, as when gale-force winds pile up a storm surge, the estuary can lose its delicate balance.

Estuaries are also especially susceptible to pollution, since their protected waters often provide natural harbors for use as seaports by man. This means an increase in dredging to keep the channel clear and open to larger ships; an increase in sewage and industrial wastes; and an increase in heat from water-cooled electric power plants.

Nature's nursery. An estuary (opposite) is sometimes flooded, sometimes dry, but it is always a place that teems with life, especially in the breeding season.

Draining the land. Wadi Hadhramaut (below) on the Arabian Peninsula displays the pattern of drainage by which rivers wash soil from the land.

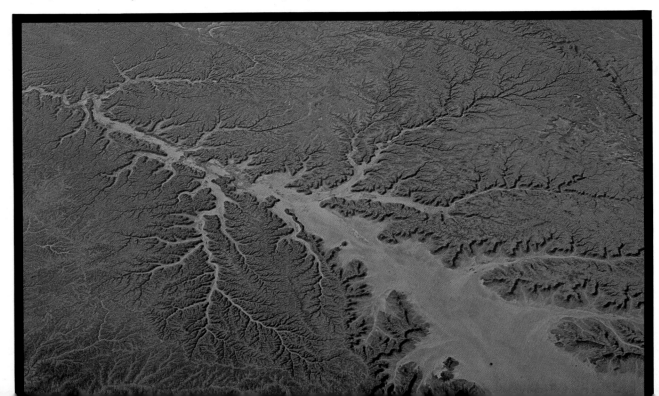

even a lobster, might be seen hiding in an underwater crevice or cave.

The distribution of these organisms largely depends on how well they can go without water and tolerate the effects of the sun at low tide. Some of the most hardy are snails and limpets which can exist on a thin film of algae they scrape off the rocks. By living in this harsh environment, they avoid many predators. On some shorelines there exists a lush growth of algae that flourishes only in the intertidal zone where it receives maximum sunlight, enough water to sustain life, and protection from grazing urchins.

Tide pools are usually colorful; in addition to the well-camouflaged brown green and gray green creatures, there are the vivid anemones and sea hares as well as the patches of worms which might be exposed tempo-

Protection. *Snails of the intertidal zone (left) huddle together in a depression to conserve moisture.*

Stranded by the Tide

A tide pool offers a microcosm of shore life, and since it is so readily accessible, a perfect place to study shore-dwelling marine creatures. The tide pools are little, or large, bodies of water left behind by receding tides. There may be numerous pools trapped by the rocky shores, or there may be only a few very broad and shallow pools, such as those found along sandier coasts. The organisms found in tide pool areas usually live immersed in water, but they must be able to withstand periodic desiccation when the tides are very low and the pools may be shallow.

With more opportunities to trap water, rocky shores provide a greater number of pools. There may be mats of seaweeds trapped in them, which provide cover for the pistol shrimp or the small fish which want to escape the foraging shorebirds. Crabs, perhaps

Zonation. *The horizontal bands of color (above) on the rocks show layering of shore organisms.*

Barnacles. *These Pacific coast barnacles (left) filter food from the water when the tide is in.*

rarily. A big orange starfish may be resting on a rock, but in the pool below there will be many more.

The sandy tide pools, less numerous because depressions in the sand are harder to come by, lack the variety of life that is found in rocky shore tide pools. But they almost always shelter small hermit crabs, some nassarius snails whose shells the hermit crabs often use for homes, sea slugs, sea hares, and small fish which were stranded when the tide went out. Because they are exposed to the hot sun with no rocks to provide shade, the sandy tide pools are often overheated reducing the oxygen content of the water, and this often results in the death of many of the organisms.

59

Building Land

Lagoons, like such other coastal features as beaches and estuaries, are temporary in nature. They indicate a transition period, as they are generally formed in areas near a specially low and fragile littoral coastline.

A lagoon generally develops where the continental shelf is smooth and broad, so that low banks or bars can build up and isolate a shallow body of water from the sea. Sometimes the formation of a lagoon is aided by the growth of coral or mangrove trees. The mangroves, which thrive in brackish water, have twisted and gnarled roots growing up from the ground as well as aerial roots extending down from the branches.

The mangrove habitat is very productive in terms of organisms, with several species of shrimp commercially fished in the lagoons

Lagoon. A lagoon is temporary and sometimes fragile and can be destroyed by a powerful storm.

and swamps. There are also the red-and-black mangrove crabs, fiddler crabs, hermit crabs, and in Indo-Pacific regions coconut crabs. In the inland areas there is the giant land crab, *Cardisoma*, which can range up to three feet across. There are also oysters, barnacles, several species of anemones, as well as isopods, shipworms, and flatworms.

Among the fish are the mudskippers, which are able to live out of the water for up to two days, and the gray snapper, which is sometimes called the mangrove snapper.

After a mangrove lagoon is formed, it continues to develop into a swamp; as it provides a basis for more vegetation, decaying organic matter collects on the bottom of the water and more tidal sediment is trapped. Only an occasional storm slows the process, as it flushes the lagoon and carries the nutrient-rich sediment out to sea.

Bay of Villefranche. *This bay on the Cote d'Azur is one of the deepest and safest shelters in the world.*

Between the Tides

As the tide rolls into shore, the tempo of activity increases. The shorebirds take their last stabs at finding food in the sand and mud, as the marine creatures begin to stir. The plankton-feeders, and the crustaceans that feed on them, are ready to dine. The acorn barnacle spreads its featherlike feeding filters at the first hint of sea spray. The anemones strengthen their holds on the substrate to make sure they don't get swept away. In the sandier regions all the burrowing animals begin preparations for leaving their homes to feed.

One trait that is common to many intertidal zone residents is the ability to hang on tightly to the rocks. In addition, they must be able to withstand the forces and abrasion of not only the waves, but also the sand and rocks the waves hurl at the shore. The abalones, snails, limpets, and chitons have protective shells and a muscular foot that creates strong suction on the substrate. The mussels secrete many hairlike byssal threads that provide firm attachment. The barnacle adheres to rocks by means of a basal plate. The soft anemones have a surprising degree of protection from the wave action. Many species cover themselves with bits and pieces of rock and shell, forming an artificial hard cover. The starfish, with its spiny skin, when pulled from a rock, will leave many of its tube feet behind still holding on. Even plants have become rubbery or calcareous, and some possess many fingerlike projections that hold the bottom tightly, forming what is called a holdfast.

The intertidal zones can offer a fantastic variety of life from one season to the next. In sandier areas there may be a population

Rocky coast. Green anemones and kelp are common sights along the Pacific when the tide is out.

explosion of bean clams, *Donax*, which inhabit the beach so densely that a bather can step on more clams than sand. Then, for unexplained reasons, they may not reappear for a number of years. The sand crab *Emerita* invades beaches by the millions at times, burrowing beneath the sand and leaving only antennae exposed to filter plankton for food.

On rocky coasts similar variations take place. Especially during the springtime, divers see great aggregations of mating snails, nudibranchs, and crabs that during the rest of the year are seldom seen. There is also a succession of organisms as the spring growth of algae precedes the increased population of grazers, which then supports predators and the eternal show continues.

Intertidal community. Starfish can be seen on the rocky shore feeding on mussels (right). At other times, they may be seen (below) among the limpets and anemones holding fast to rocks.

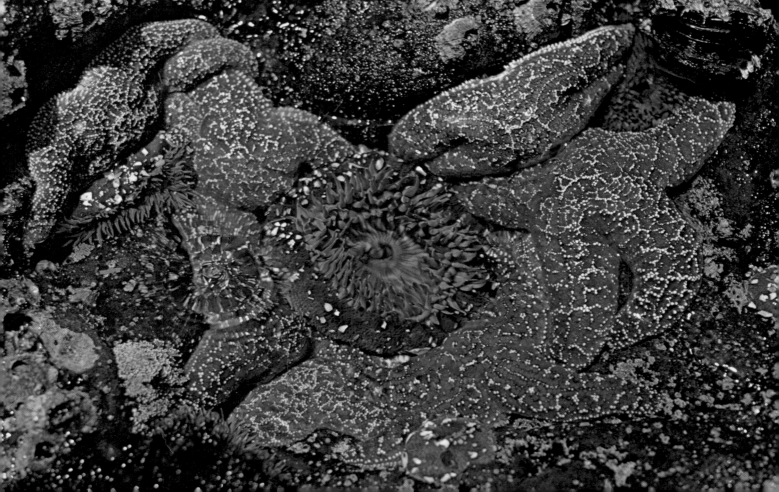

Hiding in the Weeds

In the area where the land meets the sea, seaweed and grass play a very important part. It may be an acid-tolerant plant like the needle thrush which can help sustain life where pollution would otherwise kill it off, or it might be beached seaweed which provides cover and moisture for sandhopping amphipods and isopods. In the shallow tide pools, floating plants offer protection not only from the sun's hot rays but also from land-based predators.

Further offshore, beyond the intertidal range, kelp beds can be found, attached to the bottom as far as 70 to 130 feet below the surface and reaching up like climbing vines.

Under the canopy. *Three kelp bass (left) find food and protection in a Pacific kelp bed.*

Lacy living pattern. *Hydroids (below), which are often mistaken for seaweed, are found in kelp beds.*

Nudibranch. *This white shell-less snail makes its home on a piece of kelp.*

Many organisms use the kelp in different ways, such as the fernlike bryozoans which encrust them or those whose free-floating eggs and larvae attach themselves to the fronds to begin a new stage in their lives.

The kelp beds of the Pacific Ocean off the coast of California house a complex biological community. The kelp, growing toward the surface, form protective canopies like those provided by the trees of a forest.

The kelp beds offer ideal attachment surfaces for filter-feeding organisms and small predaceous invertebrates. One study estimated that there were 100,000 organisms per square yard on some kelp fronds. These include tiny mysids, flatworms, copepods, isopods, and gammarid amphipods, as well as sessile bryozoans and anemones.

The most common fish in the kelp beds are kelp bass, kelp perch, blacksmith, and top-smelt in the higher reaches and several species of rockfish at greater depths. They feed on many of the organisms living on the kelp, especially the gammarid amphipods and other shrimp, crabs, and polycheates. A number of fish graze on other species of algae growing on the kelp.

Even though the kelp itself may not be essential for the survival of many species, the protection and associated food sources it provides enables many species to exist in the kelp ecosystem. Where kelp beds have been lost, there is a loss of animal life.

Other types of nonplanktonic plants grow in the ocean, such as eelgrass, which affords protection to pipefish and *Sargassum*, which forms a great bed in the Atlantic and is a spawning ground for Atlantic eels.

Chapter VI. Below the Deepest Tides

The relatively smooth surface of the ocean, broken by the rhythmic waves and the occasional arc of a leaping fish, actually conceals the true contour of the continents as the sea floor slopes from the landmasses toward the abyssal depths. The slope may be gentle and then steep, or it may be precipitous from land to sea, but in either case it marks the beginning of the ends of the continents.

The true outline of the landmasses has only recently been revealed, but all the maps of the world would be useless if the oceans rose or fell again as they have done so often in the past when glaciers alternately advanced and retreated from their polar domain.

We are currently in a period of rising sea level, which began about 20,000 years ago, an eyeblink in geological time. But in some areas beaches rise and fall independently of sea level. Such is the case in California, where recently formed beaches are now several feet above the water. On the East Coast, however, the irregular and sinking shoreline

---

> **"As man crowds his land and depletes its resources, he looks to the broad expanse of the ocean for help."**

---

is evidence of both past severe glaciation and a rising sea. The phenomenon is, of course, worldwide. One classicist has calculated that of the 300 important port and coastal cities built between 3000 B.C. and 476 A.D., half of them are now underwater.

The seas above the continental margin do hide the shape of the earth, but they also reveal the fantastic variety of marine life. In this region beyond the lowest tides, affected only by the deepest waves and the most violent storms, life teems. Here most species of ocean creatures are represented and here man obtains much of his seafood.

The offshore province of the sea can be subdivided into the benthos, or bottom world, and the pelagic, or open-water, region. Each is distinct: benthos includes burrowing and sessile creatures as well as motile scavengers. Above them are the nektonic animals, the swimmers that are free to range anywhere in the world. Some species, like the bluefin tuna, seem to be nearly ubiquitous, while others limit themselves to certain latitudes or smaller areas of the vast ocean. The limits might be determined by food supply or temperature or some innate and, yet poorly understood static disposition.

No environment in the sea is more diverse than that of the reef with the coral and algae combining to provide a rocklike shelter, living space, and in some cases food for countless individuals and any number of species.

As man crowds his land and depletes its resources, he looks to the broad expanse of the ocean for help. Each year more marine organisms are used to provide food for humans, and the sea floor is evermore being punctured with holes probing for oil and natural gas to satisfy an apparently unquenchable thirst for energy.

The offshore regions of the ocean are the most accessible to man, and it may be that the adage about familiarity breeding contempt will prove true.

Sampling on the shelf. A submersible's manipulator collects sediment, growth, and debris from the floor of the ocean at a depth of about 500 feet near the edge of the continental shelf in the presence of anemones, tentacled crinoids, and colorful fish.

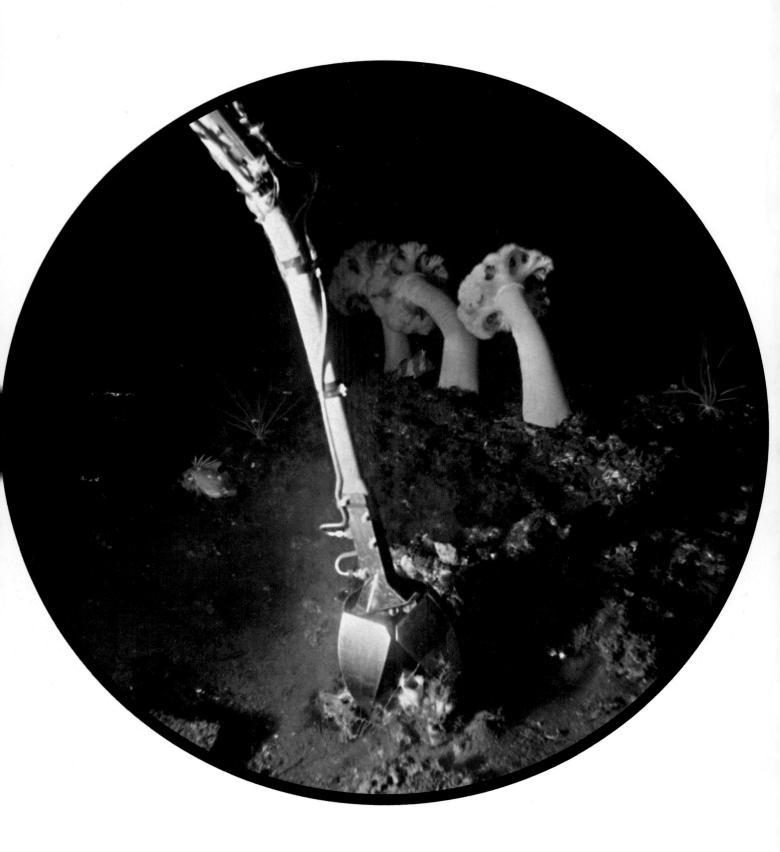

On the edge. *The color variations of the water off Cape Hatteras, North Carolina, illustrate both the sediment being carried off the continent and the different depths of water on the continental shelf.*

Finis Terrae

The outlines of the continents have been altered many times in the past and continue to undergo modifications through the action of tides, waves, rivers, and storms as well as volcanic eruptions, earthquakes, or landslides. Drastic alteration of these features has taken place through the action of glaciers. When these huge sheets of ice form, they take in great quantities of water from the ocean by keeping rain and snow from returning to the sea, and they dramatically lower the sea level. But as the heavy glaciers start growing and advancing, they depress the landmasses, and the oceans overflow their basins and flood great areas of the continents. When the glaciers melt and retreat, which they are now doing, the land recoils and rises, and the level of the sea rises. The regression of the ice and the rise of the sea level are not happening continuously; there are periods of stagnation during which the waves have carved into the cliffs formations called fossil beaches and caves. Exploration submarines have identified these phenomena all around the world, principally at minus 340 feet and minus 50 feet.

Below the water, off the shores of the continents, are the true edges of the continents. The edges may be at the end of a broad, slightly sloping plain known as the continental shelf, as occurs in the Atlantic Ocean, or the edges may be very close to the present shoreline, the continent sloping steeply from the top of a coastal mountain range to the deep ocean trenches not far offshore, as off the Pacific coast of South America. Often these underwater continental shelves are scarred and marked from past glaciers. Or they may be terraced, revealing the successive lower levels of the ocean at various periods in the geologic history of the world. In the Gulf of California, the transition is very abrupt and there is little or no trace of the continental shelf and waters get very, very deep very, very quickly.

At the edge of the continental shelf, along a line of flexion called the "kneeline," the incline becomes sharply steeper in the area known as the continental slope which drops rapidly toward the bottom. In many cases, before the ocean floor is reached, the slope flattens out again in a formation called the continental rise, which is really the buttress, the counterpart of the landmass. In some areas this break is hard to determine, as it is off the coast of California or off Florida in the Atlantic, and the area is called the continental borderland.

In the offshore area of the continental slope, deposits of sediment are usually muddy and accumulate slowly because the steep angle of the slope interacting with other forces causes the sediments to slump down to the lower regions which are level. By this means land-derived sediments reach the abyss.

Profile of the shelf. *The slope of the continents from the highest points of land, the mountain ranges, down to the abyssal plains is characterized by broad relatively flat areas separated by abrupt declines. Just beyond the coastlines lie submerged shelves of land, the continental shelf, that descend to the deep ocean floor. Upon reaching abyssal bottom, the continental slope flattens into plains, punctuated in some ocean regions by the deep oceanic trenches. Some coasts lack a continental shelf, instead nearby mountains plunge to the deep ocean bed. The continental slope is the true end of the continent.*

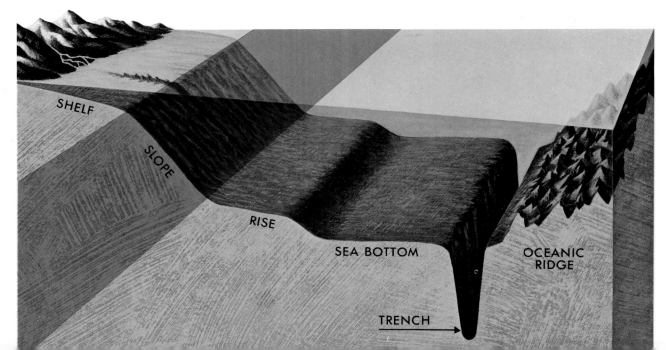

SHELF

SLOPE

RISE

SEA BOTTOM

OCEANIC RIDGE

TRENCH

Life on the Bottom

Burrowing marine organisms range from those which live in the intertidal zone, to those which spend all their lives under water in shallow waters, to those which live on the bottom and never see the light of day. These organisms which live in, or attached to, the ocean floor are called infaunal. The more mobile creatures which live just above the floor, often stirring the bottom for food, are called epifaunal. Together, they comprise the benthic, or bottom, life.

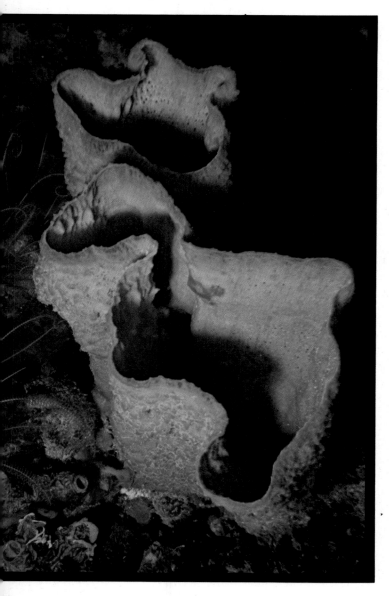

The benthic zone ranges from the deepest parts of the ocean up to the areas affected by the tides, but the most productive in terms of organisms is that region over the continental margin which is unaffected by tides. Here is found a great variety of animals from many groups including some so small they can live between grains of sand. There are worms, crustaceans, tiny starfish, and protozoa just to name a few.

The larger organisms include lugworms, surf clams, and the exotic featherdusters. These infaunal forms usually feed by extending filters into the gentle current or by straining the muddy sediment, taking whatever food they can find.

Bottom-dwellers depend for much of their food on dead and decaying organisms dropping from the shallower waters above. Many of the benthic organisms, then, are scavengers, especially among the mobile epifaunal creatures. In shallower areas, where light can penetrate to the bottom and where plants can grow at all depths, life is much more plentiful. The deeper the water, the less life there is likely to be. But as long as there is some rain of debris from above, there is some life on the bottom to take advantage of it.

Green-and-blue sponge. This sessile creature (left) leads a quiet life at the bottom of the sea, feeding on microscopic material that is pulled through its porous skeleton by its collar cells.

Kellets whelk. These gregarious scavengers (opposite, top right) explore the bottom and even each other for food. Certain times of the year hundreds of these whelks congregate for mating and afterwards lay great masses of encapsulated eggs.

Orange sponge. Another member of the phylum Porifera (opposite page, top left) displays unusual shape and color.

Standing guard. Sea pens (opposite page, bottom) remain aloof as a sea anemone flits about trying to stay out of the grasp of a predacious starfish.

Hazards of Benthic Life

The red crab, *Pleuroncodes planipes,* is both benthic and pelagic during different stages of its life and offers a study in contrast of lifestyles. As a mature adult, over two years old, the red crab is found living on the floor of the ocean near the outer edge of the continental margin off the coast of Baja, California. Younger and smaller crabs are found closer to the shore but occasionally swim far above the bottom to feed on plankton.

Great numbers of the larger crabs live in the muddy gray sand at depths ranging from 250 to 1000 feet, where they stir the bottom for food. They are rarely found on rocky and rough bottoms, where there is less of the kind of food these scavengers need.

Large numbers of the red crab are periodically found washed ashore as far north as Monterey, California, or pushed far out into the Pacific beyond their normal range. Their appearance so far from their primary distribution area is attributed to currents. When

the younger crabs near the shore rise to feed on the plankton, they may be swept to the south and west by the California and Equatorial currents or to the north by the California and Davidson countercurrents.

Not only are these crabs subjected to the currents, but when they abandon their benthic way of life, they become prey for such large free-swimming fish as the albacore, yellowfin tuna, and skipjack tuna, which normally don't range close to the bottom. Related crabs have been observed using their

Pleuroncodes planipes. *The pelagic red crab (opposite) lives a benthic life, but occasionally becomes free-swimming. Great numbers (above) then face the hazard of being swept by currents.*

swimming ability to cover considerable distances to attend huge gatherings, 1000 feet deep, at the mating season.

The benthic life, then, offers a certain amount of protection for the epifaunal organisms like crustaceans, brachiopods, nudibranchs, or sea urchins. Though the environment is harsher, isolation is protective.

Fancy Free

The waters above the continental margins are the most fertile as far as commercial fishing is concerned, for it is here that most of the world's free-swimming organisms, or nekton, can be found. Many of these creatures feed on plankton, while others prey on smaller marine organisms.

Free-swimming forms are also subject to environmental factors such as temperature and pressure, but we are not sure how much these factors influence the otherwise unrestricted range of nekton. This environment is more three-dimensional than the benthos, for the swimmers may move up, down, or in any lateral direction for great distances.

Nekton includes everything from free-swimming molluscs, like squid and octopus, to all varieties of fish, to mammals like whales, and to man with his aqualung, submarines, and bathyscaphes.

Nektonic life helps provide the organic material needed for life on the bottom, and the decomposition of dead nekton by bacteria is a source of raw material for the producers when elevated to the upper regions by upwelling currents.

Among the fish forms of nekton is agnatha, or the jawless fish, perhaps the most primitive of fishes. This class includes sea lampreys and hagfish which attach their circular sucking mouths to their prey when they attack. Sharks and rays are, of course, part of the nekton as are most of the bony fish which are found in the sea. Sea snakes and turtles are reptile representatives, while the penguin is only one of a large number of birds which depend on the sea for food.

Nekton. Free swimmers of the continental shelf area include the octopus (below), queen angelfish (opposite page, top), razorfish (opposite page center, left), lionfish (opposite page center, right), and squirrelfish (opposite page, bottom).

Living Walls

Charles Darwin, a naturalist more noted for his theory of evolution than for his work on the marine environment, speculated in 1832 on the origin of coral reefs and atolls. It was more than a century later that he was proven essentially correct. We have explained his theory and described the coral reef communities in Volume IX.

With approximately 400 true atolls and 80 million square miles of coral reefs built by the tiny polyps, these constructions qualify as the only animal-built province of the sea. It is a vast but shallow province; its richness, in opposition to benthic or polar provinces, is due to its variety. The reefs comprise very few specimens of each of many thousands of species closely intermingled. They are a center of attraction in the middle of the "blue desert" of the open ocean, providing homes for many other species of creatures. Coral reefs are some of the most exciting places on earth for the exploring diver.

The influence of the reefs in the metabolism of calcium and carbon in the sea as well as in the atmosphere is of considerable importance. Not only for their beauty and for their pathetic frailty, but also for the beneficient part they play in the concert of life, all coral reefs of the world should come under the full protection of mankind.

Coral reef community. Coral reefs provide food (opposite page, top) as well as shelter and protection for literally millions of marine creatures. The reef (opposite page, bottom) forms a barrier between the rough waters of the ocean at the top of the picture and the gentler, quiet waters of a lagoon at the bottom of the picture. Water temperature is also important, since corals are usually found where the water is between 68° and 86° F. Once the calcareous reef begins, it is subjected to erosion by the pounding waves. But bottom currents carry the sediment out to sea and prevent the reefs from becoming buried in their own debris so that colorful fish (below) of the damselfish family can still find a home. Damselfish are found in oceans throughout the world, usually in association with coral reefs, but there are numerous other types of fish (right) living in the reef environment such as jacks, wrasses, surgeonfish, blennies, and parrotfish.

Black Gold

While fishing is the most widespread commercial exploitation of the ocean, oil and gas are the most valuable commodities obtained from the sea. Chemically referred to as hydrocarbons or fossil fuel, gas and oil are organic in origin. Though there is still considerable controversy as to exactly how oil was formed, many scientists feel it originated when living creatures died and their remains accumulated on the floor of the ocean. Once there, they got trapped by an anticline or they seeped into cavities and porous rocks. Time and pressure combined to convert the organic matter to oil and gas.

In some areas large oil deposits can be found in fossile coral reefs, which offer a porous and permeable storage area for the petroleum. The Gulf of Mexico yields oil found adjacent to salt domes, which are formed when the lighter salt accumulations begin to rise through layers of denser material. These domes, called diapiric structures, are found on land and on the sea floor, especially in areas of the outer continental shelf and the continental slope. These domes may rise several hundred feet above the ocean floor.

The petroleum deposits in offshore areas are considered valuable natural resources. Oil companies lease drilling rights for hundreds of millions of dollars. There are rich petroleum deposits off the coast of California, and both oil and natural gas are being obtained in the Gulf of Mexico off Texas and Louisiana. Extensive deposits are also found off the north coast of Alaska in the Beaufort Sea, in the North Sea, and in the Atlantic Ocean 30 to 50 miles offshore, ranging from Delaware to Maine. Others are likely to be confirmed in the Antarctic and Mediterranean.

Offshore oil platform. Pictured left is one of the many complex structures built to drill for fossil fuel beneath the waters of the Gulf of Mexico.

Chapter VII. The Open Ocean

We have explained in earlier volumes of this collection that the energy for all life comes ultimately from the sun. Plants have the ability to obtain this energy directly through photosynthesis, but animals must get it secondhand by eating plants. For many years it was assumed that the only vegetable life in the ocean was the seaweed attached to the bottom or floating in places like the Sargasso Sea or the algae that covered the surfaces in well-illuminated shore waters.

It wasn't until early in the last century, long after the discovery of the microscope, that plankton was discovered. Men finally began to learn of the wondrous variety of minute plant and animal life in the sea.

The plant plankton, or phytoplankton, when they are near the surface, can utilize the sun's rays in photosynthesis. But phytoplankton

---

> **"Sharks, rays, tuna, and marlin are among the fish that favor the wide open spaces."**

---

are not rooted to land and float about in the ocean, providing food for both zooplankton and larger organisms. The average productivity of an acre of open ocean is certainly inferior to that of rich soil, but higher than that of a desert. And, though this productivity may vary greatly, it is far more constant all over the seas than land's output all over the world.

Not all plankton is microscopic, since such visible creatures as jellyfish qualify as free-floaters. And not all plankton float for their entire life. The permanent plankton is called holoplankton, while temporary plankton is called meroplankton and might include the larvae of pelagic or benthic organisms. It is this stage where the young are most vulner-

able, they are decimated by predators both large and small. Fish may filter them from the water on fine gill rakers. Tiny medusas may ensnare them in stinging tentacles or voracious little arrow worms may impale them in bristling jaws.

In addition to plankton, the open ocean also feeds some very large creatures. Sharks, rays, tuna, and marlin are among the fish that favor the wide open spaces. Many of them are very fast swimmers and prey on squids and smaller schooling fish like the flyingfish and halfbeaks. Sea mammals and whales also wander the ocean world.

The productivity of this region of the sea is limited because the nutrients are in short supply and without this fertilizer the floating microscopic plants cannot populate in the vast numbers seen in coastal waters. For the investigator, the open ocean is like a cobalt blue desert dotted by occasional concentrations of creatures, large or small. In spite of this relative paucity of life, there exist predator-prey relationships like those in coastal waters. A herring may feed upon arrow worms, copepods, shrimp, and tiny fish. The tiny fish may also prey upon copepods, while the copepods prey on diatoms. Jellyfish prey on many of the same organisms as the adult herring and even eat the young herring, but they do not serve as food for the adults. These interlocking feeding relationships are called food webs, as contrasted to the direct one-to-one situation of a food chain. Actually, very few food chains exist, for the sea is as complex as life itself.

Lookdown fish. This Atlantic Ocean swimmer of the genus Selene *is popularly called the lookdown fish because its eyes are placed in the head so that it often appears to be looking down while swimming straight ahead.*

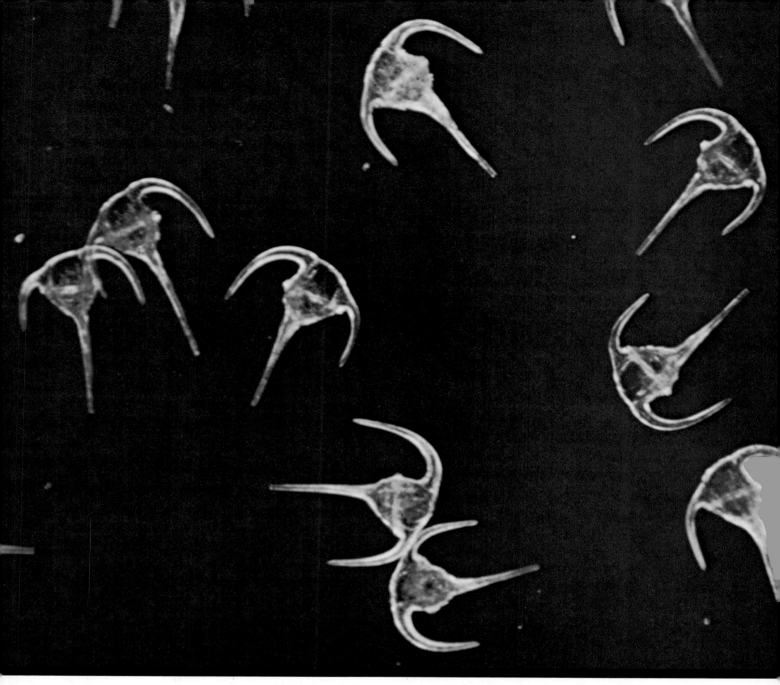

Start of Something Big

Phytoplankton, those tiny free-floating plants that directly or indirectly provide energy from the sun to so many ocean creatures, are mostly water. Over 80 percent of their weight is water, with the rest of their tiny mass consisting of protein, fat, carbohydrates, and mineral components—either calcium or silica compounds—making up their shells or skeletons.

Wherever there is radiant light to trigger photosynthesis, dissolved nutrient salts, and carbon dioxide, that is where phytoplankton can blossom. And nearby are the creatures that feed on microflora. The sun penetrates in sufficient quantity for photosynthesis to perhaps 300 feet deep in the ocean, and this defines the photic zone. The higher up in the photic zone, the more sunlight there is, and as a result, the more productive the area. But in some still tropical waters where sunshine

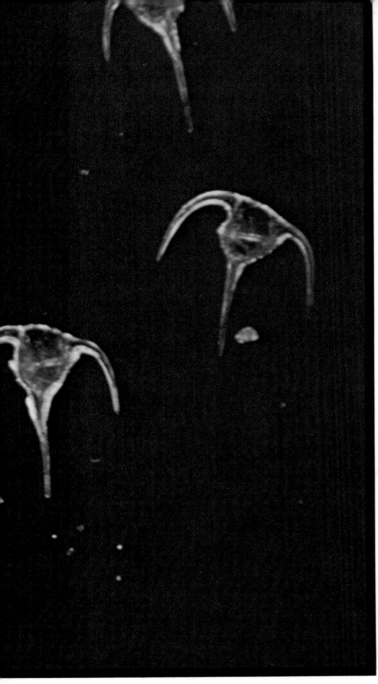

they can multiply so fast, doubling their number in 48 hours, that if they were unchecked by plankton-feeders they would clog the seas in a matter of weeks. The name diatom—meaning two atoms—is derived from the fact that each cell wall consists of two nearly equal overlapping halves. All diatoms are single-celled, but some reproduce and form complex chains.

Diatoms have a hard, glassy shell made of silica, which is pitted and grooved. When these plants die, their shells litter the ocean bottom and eventually become part of the sedimentary rock being formed. This rock layer is called diatomaceous earth and is used commercially for making glass.

Coccolithophorids and dinoflagellates are the two other major types of phytoplankton. The coccolithophorids are distinguished by the calcareous plates that cover their external surface. Dinoflagellates are the most complex of the phytoplankton. Some may have traces of chlorophyll and others do not, indicating they may be part animal as well as vegetable. One factor that enables dinoflagellates to succeed is their motility. With their two flagella, they are able to propel themselves vertically. They have been found to descend below a shallow thermocline to absorb nutrients at night, then ascend to the surface for light during the day. Consequently they are able to bring nutrients to the sunlight and carry on photosynthesis in a nutrient-poor environment. These phototropic vertical voyages are exactly the opposite of the general migrations of the "deep scattering layers," when billions of tons of creatures sink during the daytime and surface at night.

is abundant, the phytoplankton may not be very plentiful because the calm water indicates there is little stirring and mixing which is absolutely necessary to bring nutrients to the surface.

Probably the most common algal plankton is the diatom, which seems to explode in numbers during the summer months in temperate and colder waters when the sunlight is present for longer periods than usual. There are so many diatoms in the ocean and

Whip propulsion. Dinoflagellates, microscopic planktonic animals, use their whiplike extensions of cyctoplasm to move vertically in the water.

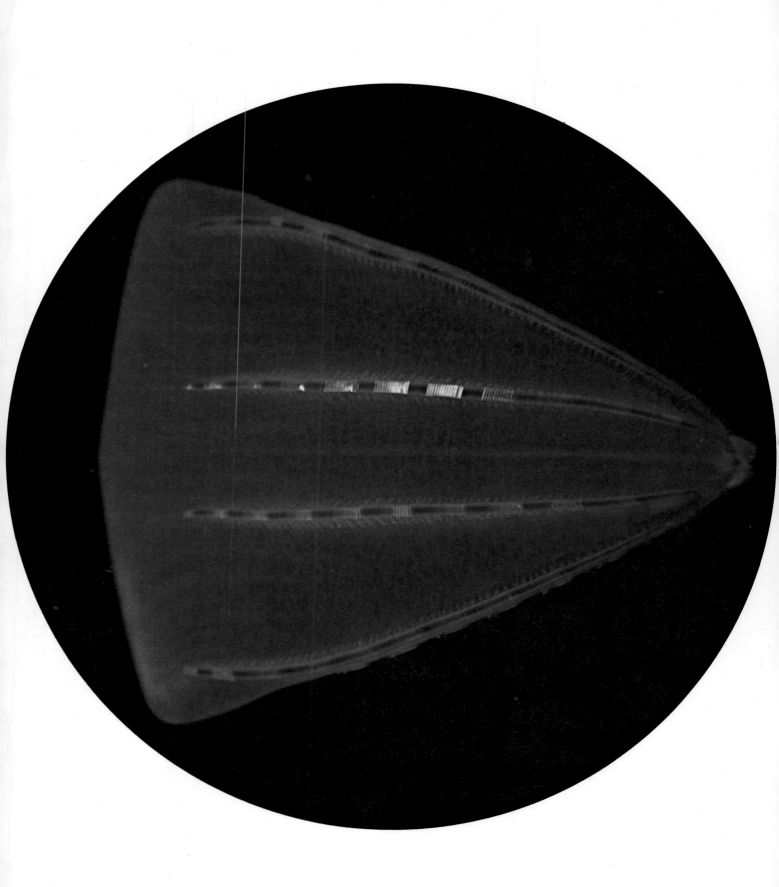

Drifting Animals

Just as there are plants that live a floating life, so are there planktonic animals, or zooplankton. In most cases, they feed upon the phytoplankton and are themselves fed upon by larger organisms. Many of the zooplankton are protozoans, or single-celled animals, and are microscopic, but others are quite large, such as the salps. Most planktonic species possess some mechanisms for propelling themelves through the water, but these are usually very weak and the organisms are subjected to current and wind.

As with phytoplankton, zooplankton contribute skeletal matter and debris to the sedimentary layers on the ocean floor. One order, Foraminifera, is used by geologists in determining the history of rock layers and the climate in which it was laid down. Foraminifera are usually snail-shaped with shells made of calcium, while the closely related Radiolaria have shells made of silica.

Among the other types of zooplankton are the larvae of larger organisms, such as crabs and lobsters; several small forms of creatures that have pelagic or sessile relatives; arrow worms, which have fins to aid them in moving through the water, jawlike bristles, and teeth for eating; comb jellies like the sea gooseberry that capture food with sticky tentacles; and the Portuguese man-of-war, a siphonophore, which is a colony of specially adapted individuals living as one adult. Crustaceans are an important part of the zooplankton. These include ostracods, various kinds of shrimp, and copepods.

Wide variety. Plankton includes diverse life, such as the ctenophore (opposite page) and diatoms, copepods, and various larval forms (below).

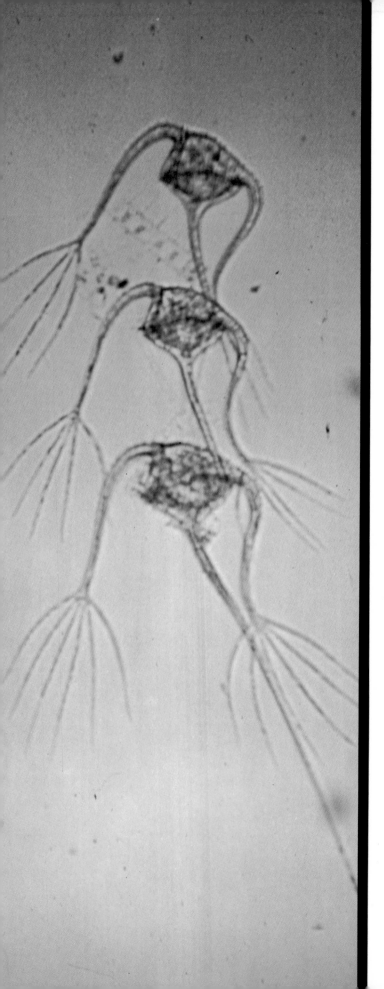

To Lead a Floating Life

With gravity constantly attracting plankton toward the bottom, these tiny plants and animals have developed features which retard sinking. An important adaptation of phytoplankton is the extension of body surface into a variety of bizarre forms that result in a frictional resistance to sinking. The diatom variations include long narrow forms, flattened shapes, the development of bristles and spines, and the formation of long chains that grow into spiral patterns. Diatoms also use various methods of chemically reducing cell density. Some animals secrete fatty globules which have no purpose other than to increase their buoyancy. Some jellyfish-like creatures (Velella, Physalia) possess sails which enable them to drift through the water while their tentacles are extended below in search of food.

The radiolarians, and there are more than 4000 species of them, not only secrete a silicate shell but also develop extensions that give them the appearance of snowflakes. These extensions are fine threads of cytoplasm that are also used for feeding.

Phytoplankton are most abundant in the upper layers of the water where light can best penetrate. But there are some diatoms, dinoflagellates, and coccolithophorids which live in the lower reaches of the photic zone. They maintain their level through such devices as the winglike parachute of the diatom *Planktioniella sol*. Some types of dinoflagellates are able to rise and sink in the water, but because of their susceptibility to currents, they are still true plankton.

Dinoflagellates. *These single-celled planktonic animals (left) use their whiplike extensions to propel themselves weakly through the ocean.*

Starting out in life. *Echinoderm larva (opposite page, top) and phornid larva (opposite page, bottom) lead a planktonic existence early in their lives.*

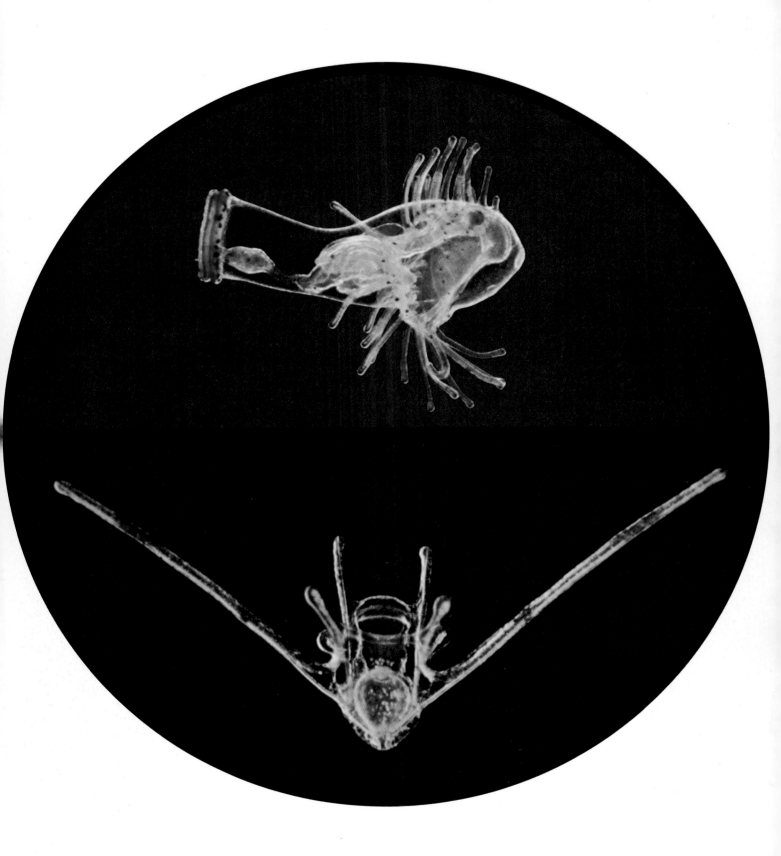

Wide Open Spaces

Speed. If one word could be used to describe the dominant swimmers of the open ocean it would be speed. The foremost residents of the area are the tunas, sharks, billfish, and dolphins. They all have streamlined bodies and the ability to produce short bursts of very high speed. Some sailfish (see Volume IV) have reportedly been timed at 70 miles an hour. And "speed kills," for these fish use their speed to hunt down slower fish sharing their environment.

Hiding places and cover are unknown to the smaller creatures of the open sea so they must develop other defenses. Schooling, although not solely a defense mechanism, must be successful since more than 4000 species exhibit this behavior. When small fish in school formations are attacked, they often crowd more closely together in what appear to be spherical patterns, looking like a constantly squirming ball made up of thousands of individual fishes. Schooling cannot overcome the speed of some predators, but the mass of fishes may present a deterrent.

There are many thousands of species of sizeable creatures in the open sea, but some of them have been far more successful than others. The basic food for the giants is essentially made of sprats, sardines, flyingfish, squid, and cuttlefish. And though they are relentlessly hunted and caught, they still thrive in great numbers. The cunning game of life and death between a dolphinfish (*Coryphaena*) and a flyingfish is astounding. The flyingfish changes course during its flight to fool the *Coryphaena,* but the hunter knows that in order not to lose track of its prey, it must always swim just under the flyingfish, and if it successfully does so, it will be there to swallow the winged fish when it ultimately must reenter the water.

Squids generally stay very deep during the daytime where only the mammals dare to chase them. But at night the squids have to come closer to the surface to feed (also upon the poor flyingfish that can know no sleep). Cuttlefish were believed to be shore animals but they are also pelagic, and recently huge concentrations of them have been spotted at a depth of 1500 feet in the mid-Atlantic. Pilot whales and orcas feed on them.

Very little is known of the pelagic life of the large sea turtles, and especially of the huge but rare leather turtle. Nothing is definitely known about where the basking shark spends most of the year, and the daily behavior of the Lords of the Sea, because of their versatility and because of man's clumsiness in the sea, remains a mystery.

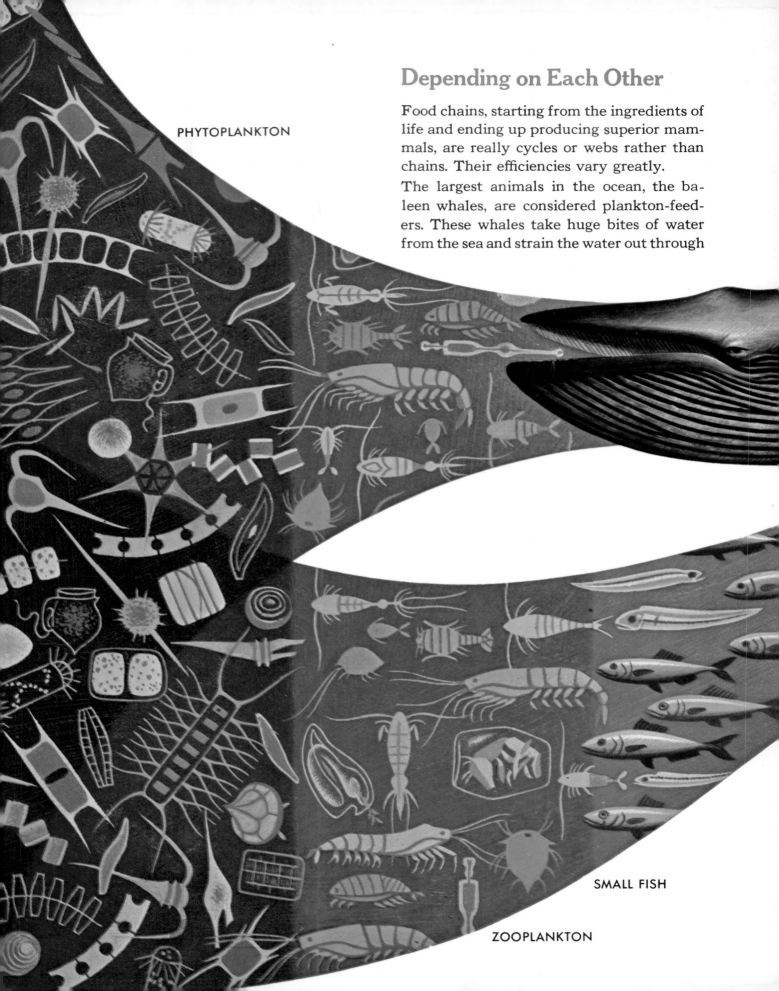

PHYTOPLANKTON

Depending on Each Other

Food chains, starting from the ingredients of life and ending up producing superior mammals, are really cycles or webs rather than chains. Their efficiencies vary greatly.

The largest animals in the ocean, the baleen whales, are considered plankton-feeders. These whales take huge bites of water from the sea and strain the water out through

SMALL FISH

ZOOPLANKTON

the bristlelike baleen inside their mouths, keeping and swallowing the krill, crustaceans known as euphausiids, which they feed upon. Other plankton-feeders have much finer filtering systems and are able to trap the tiniest particles, even the nanoplankton, or dwarf plankton. While many fish feed on zooplankton, there is probably no fish which lives exclusively on phytoplankton, unless it is the milkfish, *Chano chanos.*

fish, and so on up the line so that it may take as much as 1 million pounds of diatoms to build one pound of shark. By contrast, only 50 pounds of diatoms provide food for 5 pounds of shrimp, and that is enough to construct one pound of whale.

The most dramatic relationship in the food web is the one that binds a predator and its prey, as each has developed more efficient methods of attack and defense over long

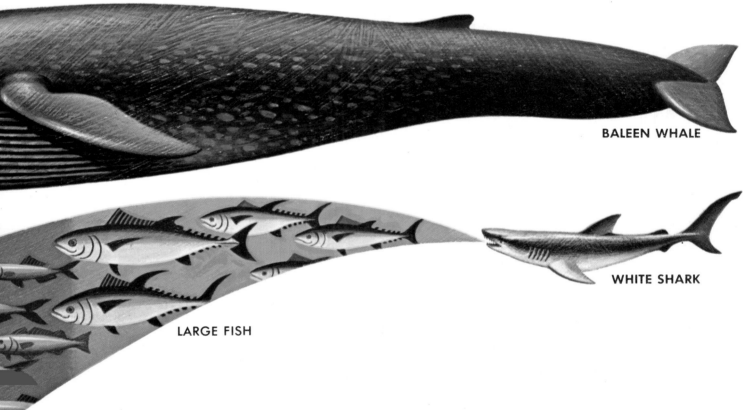

BALEEN WHALE

WHITE SHARK

LARGE FISH

Also part of the cycle are the scavengers that roam on or above the bottom in search of decaying organic material which may have come from above. Scavengers also eat algae, bacteria, and zooplankton as they forage through the sediment.

The transfer of organic material through the cycle is more or less efficient: it takes about 10 pounds of phytoplankton to produce a pound of zooplankton, and about 10 pounds of zooplankton to produce a pound of flying-

By feeding nearer the source of the food web, the ***baleen whale*** *has an exceptionally high efficiency ratio as a marine protein builder. In contrast, the* ***white shark*** *has a much smaller amount of food available and a low economic potential.*

periods of evolutionary time. Those that do not play the game well fall by the wayside. Sometimes the predator is so small in relation to its victim that it has a special name: parasite. These organisms not only feed off their victims but also live on or in them.

Here, There, Not Everywhere

Nothing in the ocean is static, and this is as true of the distribution of plankton as it is of temperature or salinity. A distinction can be made, for example, between the near-shore zooplankton which contains a large number of larvae of benthic animals, while further offshore, there are only the larvae of nektonic or planktonic creatures. A convenient, but by no means absolute, dividing line is the 600-foot mark, which roughly corresponds to the edge of the continental shelf. Environmental factors have such an influence on organisms that many will vary their habitat from year to year, season to season, or even day to day because of them. The great transoceanic migrations will be considered in a future volume, but there are also vertical migrations not only of plankton but also of larger organisms. The reaction of animals to light is far from being uniform; zoo-

*These **two recordings** from an echosounder show the profile of the abyssal plain and the approach of the Atlantic Ridge on the 3000- to 6000-meter scale as well as several layers of life reflecting sound be-low the surface. Both were taken just before sunset and indicate that some of the layers remain at 300 and 500 meters while several others rise from 500 meters toward the surface.*

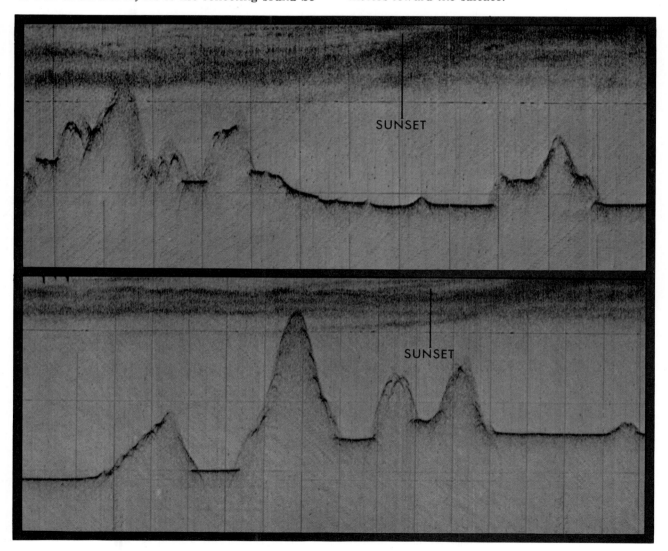

plankton and small fish like anchovies or sardines are attracted by a lamp; larger predators are attracted by the presence of the smaller ones but fear the light and stay on the outskirts of the illuminated zone.

Most fishermen know that some creatures can be caught near the surface at night, but that during the day nets must be lowered several hundred feet to obtain the same animals. It is assumed that these organisms were seeking protection in the poorly lit waters below and ascend under the cover of darkness to feed on life forms above.

This vertical migration causes the "false bottom" effect on echo-sounding devices, for when the pulses are sent out during the day, they are reflected back far short of the true bottom. This deep scattering layer, DSL as it is called, may disappear at night or may be seen to rise to the surface. Some organisms making up these layers are schools of small hatchetfish, crustaceans, and even the si-

phonophores, which are jellyfishlike creatures. But mystery still surrounds many aspects of the phantom layers.

There are also patchwork patterns in the distribution of marine animals as a result of the schooling behavior of pelagic fish.

For thousands of reasons life is constantly redistributed throughout the oceans; a few areas remain highly productive, while others qualify as virtual water deserts.

*This **12-kilocycle echogram** shows the beginning of the daily vertical migrations of plankton creatures in the open sea as revealed by the so-called "deep scattering layers." Before dawn, at approximately 4:30* A.M.*, there are three layers near the surface. At 5:00* A.M. *the morning light begins to filter through the water and the uppermost layer divides into various sloping lines indicating the descent of individual populations. With the rising sun, as many as six such different populations are shown seeking the protection of darkness below and sinking at various speeds. The first creatures to decend were the fastest and did so at one half-foot per second!*

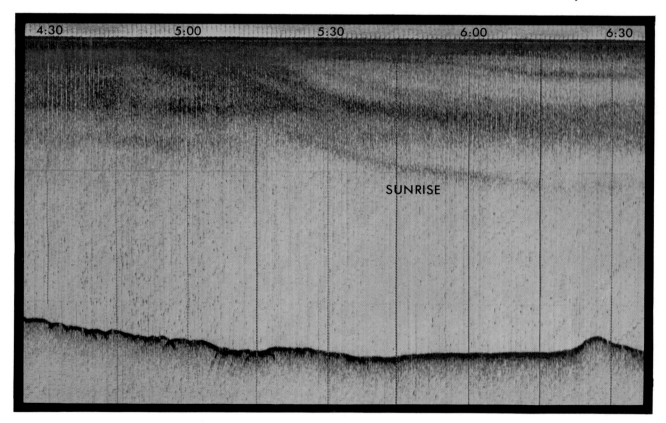

SUNRISE

Chapter VIII. Facing the Abyss

The oceans are deep, very deep, but far below the last glimmer of sunlight life goes on. In some cases the organisms are bioluminescent, they provide their own light, like the lanternfish. There must be special adaptations in a world with no light, no green vegetation, and low water temperature.

The deep sea, or abyssal region, is described as the single largest environment on earth, since it covers 85 percent of the areas. we know as ocean basins. For vast reaches the abyss is a smooth plain, covered with thick

---

"The turbidity current may flow like a river on land, carving its bed in the almost flat sediment of the plain."

---

sediments, much of it moved by turbidity currents that occasionally rush down gorges in the continental slope called submarine canyons. The sediments along canyons are stirred up by the current and slide, gaining momentum and more sediment as the mass proceeds all the way down to the abyssal plain. There the turbidity current flows like a river on land, carving its bed in the almost flat sediment of the plain. When the velocity begins to slow down, the heavier particles begin to settle out. In a process known as graded bedding, the particles get finer and finer further from land.

Occasionally the abyssal plains are broken by volcanic mountain peaks, which periodically protrude above the surface to form an island. More often the peaks are totally submerged and are called seamounts. Some of these seamounts are flat-topped, having been leveled, perhaps by wave action when they encroached upon shallow waters, and they are called guyots.

Probably the most prominent feature of the underwater profile is the midocean ridge, a mountain range which doesn't exactly ring the globe, but which does extend from the Atlantic through the Indian Ocean and to the Pacific, where it ends in close proximity to the American continents.

Life in the abyssal zone has its pelagic types, like the squid, rabbitfish, deepwater cod, anglerfish, rays, and eels. But life is mostly concentrated on the bottom. The sessile and burrowing creatures, the crawling and weak-swimming organisms all compete for the organic material—living or dead—which comes their way from above or is brought by currents. Much of their activity takes place in sediments whether turbidite or contourite. Unlike turbidite, carried downhill by turbidity currents, contourite is terrigenous mud moved along the bottom by currents that parallel bathymetric contours.

Life in the ocean contributes its share to the deposits on the bottom, whether it is the whitish yellow or brown chalky Globigerina ooze, the straw or cream colored diatom ooze, or the pale greenish yellow radiolarian ooze. The siliceous and calcareous skeletal material of these plants and animals falls to the bottom, sometimes playing a determinate role in the type of rock layer which will form in that area of the seabed.

Another process taking place in the deep sea is the precipitation of metals like manganese, iron, and cobalt, which coat rocks and organic debris. Deposits of these metals are widespread in some areas and are attracting the attention of underwater miners.

Low-down life. This shark of the deep has been lured to the camera by the aroma of food. Notice this shark lacks a dorsal fin.

The French bathyscaphe Archimede *gave man his first in-person look at the abyssal mid-Atlantic ridge. The three-man vessel can dive as deep as 33,000 feet beneath the sea.*

The Deep Plains

Beyond the reaches of the continental slope lies the abyssal plains, the vast, often smooth and flat region described on page 94.

The abyssal plains cover twice as much area as all the exposed continental land in the world. In most cases, these plains are not subject to the same wrinkling forces continents are. Another reason the abyssal plains are so vast and smooth is sedimentation. Rather than having been scoured smooth by

wave action or leveled by glacial action, the original surface of the plains was hard and irregular. These irregularities were then filled in by sediment swept off the continental slope, and the surfaces of the plains were evened out, in much the same way new-fallen and windblown snow evens out the profile of a rock-strewn field.

The environment of the abyss is far different than that of the near-shore areas and the surface layers. Below 6000 feet, in total obscurity, there are no plants. The salinity of the water is an almost constant 35 parts per 1000, with the temperature of the water usually averaging around 36° F. The temperature of the bottom itself does vary, though, ranging from a high of about 55° in the Mediterranean to a low of about 28° in the polar regions. There is accordingly a constant exchange of heat between deep water and the subsoil.

The abyssal environment is not totally uniform with regard to topography. In addition to the occasional mountain peak, there are abyssal hills, which are perhaps most prom-

Profile features. The major portion of the ocean floor consists of flat expanses known as the abyssal plains extending between the margins of the continents and the midocean ridges.

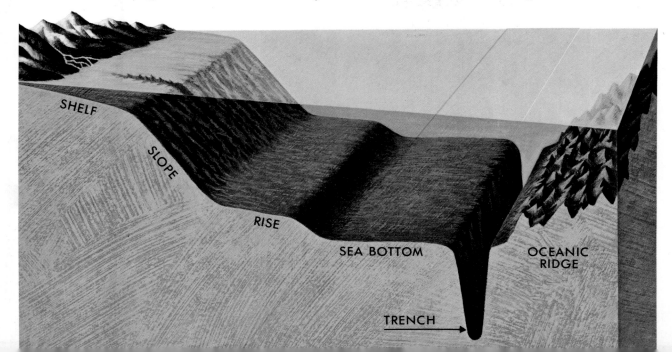

SHELF

SLOPE

RISE

SEA BOTTOM

OCEANIC RIDGE

TRENCH

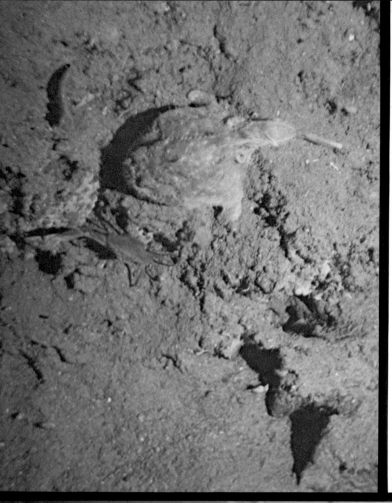

inent in the Pacific Ocean. The Pacific is not as thickly sedimented as the Atlantic, part of the reason being that more of the large, sediment-laden rivers of the world empty into the Atlantic and as a result much of the topography of the Pacific has not been as smoothed over. This partially helps explain why more abyssal hills are detected in the Pacific. They may be several miles wide and range anywhere from 100 to 3000 feet above the floor of the ocean.

Their origin is still something of a mystery. Some scientists believe abyssal hills occur as a result of volcanic action, while others feel they may be of sedimentary origin. Thanks to deep coring and exploration submersibles, we will soon know.

Life in the deep. *There are some strange forms of animals in the deep of the sea, such as this batfish (left), pictured with a more familiar-looking crab, and sea cucumbers and brittle stars (below).*

A Striking Profile

One of the more outstanding features under the sea are the seamounts. They may occur as isolated mountains, or they may poke up through the water as visible islands, such as the Cape Verde Islands. These are volcanic in origin, as are all seamounts, and it is not surprising to find that there are many seamounts in the Pacific since it is an area of great volcanic activity.

Volcanic seascape. *Stark basaltic rocks give the look of a moonscape rather than the bottom of the ocean. Some islands even have black sand beaches that were formed from volcanic rocks.*

Ordinarily the term seamount is not applied to peaks that extend above the water, like the Hawaiian Islands, but only to underwater mountains, like Great Meteor Seamount in the northeastern Atlantic. This totally submerged peak rises 13,000 feet above the sea floor and is 68 miles in diameter at its base, but we will see that it is improperly named a seamount.

As these volcanic seamounts push their way up from the seabed, they may eventually get close enough to the surface for coral to begin reef building, provided, of course, the surface temperature of the water is appropriate. If the seamount does not extend much above sea level, the action of the waves can level the peak. When the seamount begins to subside, the coral is eventually pulled below the limits it can survive and reef-building activity ceases. These flat-topped volcanic mountains, often associated with coral reef flora and fauna, are called guyots. In profile they look like an underwater truncated cone. This is the exact case of the so-called Meteor Seamount. One of the largest guyots measured, in the Pacific, is 35 miles across its flat top. The height of guyots, which by definition must be totally submerged, ranges from 3000 to 5100 feet below sea level.

Another feature on the profile of the ocean floor is the midocean ridge system. This is a continuous chain of mountains that stays beneath the surface for the most part and extends down the middle of the Atlantic, around the tip of Africa, and splits as it heads east. One spur travels up the axis of the Indian Ocean, while the other continues east through the Pacific Ocean until it approaches the coast of South America, where it splits again. One branch continues almost up to the South American continent itself, while the other heads north until it ends off the coast of California. In the Pacific, the name midocean ridge system is misleading, since it is located to the south and to the east of the central axes of the ocean. The ridge is split all along its extension by a very deep and narrow rift valley.

In some places the midocean ridge extends above the surface to form islands, such as

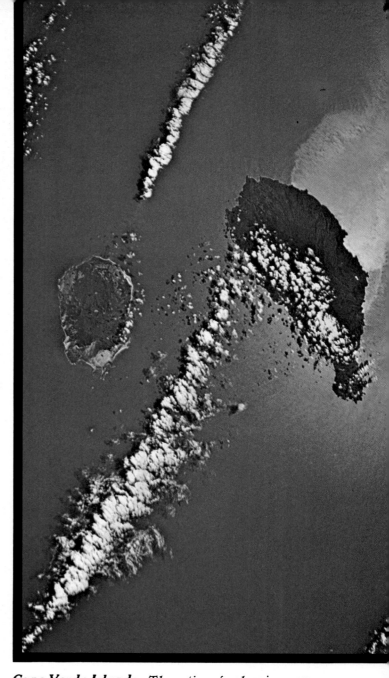

Cape Verde Islands. *These tips of volcanic mountains affect both the wind and surface currents, which turn right to flow around the islands. The light spot at the top is a calm area of upwelling.*

the Azores and Tristan da Cunha. Branches and spurs of the system are high enough to affect deep-water circulation. The Walvis Ridge in the southeast Atlantic, for example, prevents the cold Antarctic Deep Water from flowing under the warmer and less dense water of the South Atlantic.

Underwater Happenings

The year was 1929 and transoceanic communication was limited to submarine cables. There was no worldwide microwave relay network, no communication satellite orbiting the earth. The transatlantic cables were occasionally broken, by whatever cause, and it was considered one of the hazards of the business. But in 1929, at about the same time an earthquake occurred off the coast of Newfoundland, several of the transatlantic cables were snapped. It was assumed that these breaks were caused by the earthquake. But later, scientists studying the records of the quake pointed out that the cables were snapped one by one, rather than simultaneously, indicating that it was probably not the moving earth that caused their rupture.

The scientists proposed that the earthquake had touched off a turbidity current of catastrophic magnitude, which had roared down the continental slope, picking up loose sediment and combining the destructive forces of an underwater landslide and of a tsunami. This rampaging turbidity current, they said, was responsible for snapping the cables. And they could tell exactly how fast it had moved; they calculated the speed by correlating the time at which service was inter-

Underwater rivers of sand. Two views of the same phenomenon off Cape San Lucas in Baja, California show how erosion continuously feeds sediments to the abyssal plains.

rupted on each successive cable and the distance between the cables. They found amazing velocities of 40 to 50 miles per hour.

Turbidity currents are usually channeled through submarine canyons carved along the edges of the continents, in which sediment slowly accumulates. When a turbidity current scours the canyon, the sediment is funneled out to the abyssal plain. The presence of these canyons, the existence of turbidity currents, and their role in distributing terrigenous sediments on the abyssal plain is fairly much agreed upon.

Some submarine canyons are located at the mouths of great rivers, like the Congo Submarine Canyon off the west coast of Africa or the Hudson Canyon off the East Coast of the United States. This suggests that the old erosive force of the river had cut the canyon when the sea level was lower. The problem is that in many cases they are far below any known prehistoric sea level. Some scientists argue that the submarine canyons are not only swept clean by turbidity currents but were actually formed by them.

101

The Making of an Ooze

The finest sediments of the deep sea have their origins on land, in the sea itself, and even from outer space. The movement of sediments off the continental shelf to the abyss is one major source of fine particles, but this material is generally limited to the margins of the abyssal plains. Another source is the ash of active volcanoes spewn into the air and distributed by the winds. The result of a period of active vulcanism may be a thin layer of volcanic sediment. Millions of years later, when overlain with other sediments, this layer may tell inquiring scientists of past volcanic catastrophes.

Our earth is constantly being bombarded from space. Tiny meteorites and minute particles of cosmic dust are pulled toward the earth and land here, just as may happen in the formation of a star, but on an infinitely smaller scale. The sea is the best place to look for these particles of cosmic dust because they are relatively undisturbed and relatively small amounts of other sediments —at least compared to areas of land—interfere with their detection.

Still another source of sediment is the ocean's plant and animal life. As they die, or are killed, various parts of their shells and skele-tons sink to the bottom. Some of the calcareous material dissolves en route to the bottom, the rest settles and collects there. Around some tropical islands there are vast deposits of organic calcium carbonate which was first deposited by plants and animals near the islands, then carried further out to sea by slumping and turbidity currents.

Since diatoms are so widespread and plentiful in their planktonic existence, it is not surprising to find that there are sections of the ocean bottom that are littered with their silica shells, in some cases many feet thick. If left undisturbed, these shells are compressed and transformed into layers of diatomaceous earth.

The shells of zooplanktonic creatures also fall to the bottom to create oozes similar to those of phytoplanktonic origin. The Foraminifera contribute calcium carbonate, while the Radiolaria add to the siliceous ooze. The shells of the unicellular Globigerinae, of the

From bottom to top. Biogenous oozes are laid down on the sea floor when planktonic animals and plants die and their shells and skeletons sink. Sometimes these sedimentary beds, after being compressed and solidified, are thrust up above the surface to form terrestrial features like the white chalk cliffs of Dover, England (opposite page), which are made up in large part of the calcareous remains of foraminifers (below) and coccolithophores.

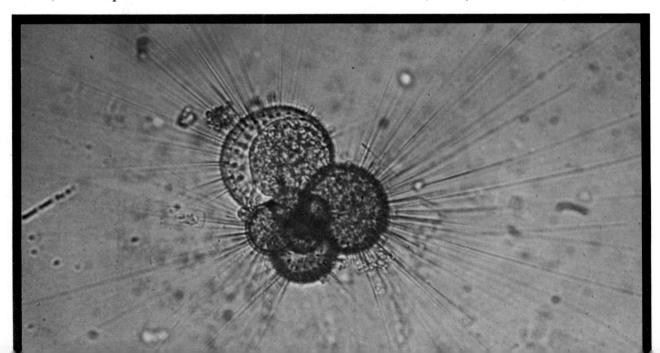

class Foraminifera, are prominent in the Atlantic, so much so that Globigerina ooze is used to distinguish areas of the abyssal bottom, often near the edges.

These oozes, also called biogenous deposits, must by definition be at least 30 percent skeletal material. There may also be inorganic material, such as red or brown clay. The Globigerina ooze is only one type of ooze named for the organism that is the most prominent in the deposit. Other types include foraminiferal, pteropod, coccolith,

and radiolarian. When one family is more prevalent, such as the Globigerinae, then the ooze is named for it.

Oozes are usually found in areas of high biological productivity or where other sediments are deposited very slowly. As a general rule, calcareous oozes are found in fairly shallow waters. This is because the rate of solubility increases with depth and lower temperatures, resulting in the dissolving of calcareous particles before they can build up into sedimentary layers.

Mined from the depths. A manganese nodule dredged from the bottom looks like a rock baseball.

Interior. Manganese laid down in concentric layers can be seen in this broken specimen.

Nugget Fields

One of the potentially most important deposits on the sea floor is manganese. This metal is precipitated in rounded nodules whose cross-sections show growth rings, like the cross-sections of trees. The manganese, which is occasionally laced with calcium carbonate, is an authigenic deposit, which means it was formed under water rather than having been carried there by currents.

The manganese is precipitated from seawater. It usually collects around some object like a fish bone or shark's tooth. The nodule stops forming when it is covered with sediment. Thus manganese nodules are most commonly found in areas where there is little or slow sedimentation or where there are strong currents which sweep the bottom of sediment. One such place is the Blake Plateau off the Atlantic coast of Florida, where the fast-moving Gulf Stream removes ooze.

Down 20,000 feet. A rattail scurries on the bottom that is strewn with manganese nodules.

Cross-section. *Cut in half, a manganese nodule resembles a cross-section of a trees growth rings.*

The floor of the Pacific, with its slower rate of sedimentation, is strewn with manganese nodules, ranging from less than an inch in diameter to the size of a baseball.

The fact that these nodules are often very nearly concentric indicates that they must be rolled over frequently to allow even accumulation on all sides. If they just lay on the bottom undisturbed, they would form mounds instead. The nodules are pushed about by the currents, very likely the same currents that bring additional manganese and iron to feed the nodules.

The largest deposits of manganese nodules are found deep under areas of low biological productivity and where the bottom waters are extremely cold. Thus they are plentiful in the abyssal plains between the polar and equatorial regions and away from the rapidly accumulating terrigenous sediments.

The origin of the nodules is not certain, but they probably occur in water which is supersaturated with manganese. As more of the dissolved element is brought into the area, some is precipitated in a manner similar to the way calcium carbonate is precipitated from water to form limestone. The layers of manganese are only a very thin film, one atom thick, alternating with equally thin layers of other elements in a process which may take hundreds of thousands of years.

The nodules contain many other elements such as iron, silicon, aluminum, sodium, nickel, and magnesium and occasionally such rare metals as vanadium, zirconium, and titanium. The manganese is usually present in the form of manganese dioxide.

Gathering place. *Manganese will form on almost any object from a shark's tooth to a swordfish's bill.*

Life in Seclusion

The out-of-sight, below-the-surface life of the abyssal zone must cope with incredible constancy. There are no seasonal currents, warming trends, longer or shorter day lengths, storms, or even seasonal migrating schools of fish. All is black and cold. The increased pressure of the lower depths is not an excluding factor for life, since most organisms are made up largely of water, which is almost totally incompressible. Simple creatures like starfish have been com-

pressed slowly in laboratory pressure tanks and have proved capable of living well in pressures equivalent to a depth of 15,000 feet. Beyond that starfish have displayed a sharp decrease in their activity.

Two of the most frequently encountered open water fish are lanternfish (Myctophidae) and hatchetfish (Sternoptychidae). Both groups grow only to a few inches in length and possess bioluminescent photophores along their sides and ventral surfaces. Squid are also frequently seen in dense schools. Nearer the bottom, rattail fish are commonly found. These chimeras, as they are also called, possess a ventral beaklike mouth and a long pointed tail and are the most primitive of the cartilagenous fishes,

making them relatives of sharks and rays. They are bottom feeders, using their mouths, which can be protruded, to shovel through the mud. They swim head down and are aided by an unbranched caudal fin which acts much like a shark's in giving a lift to the posterior, causing the head to point down.

Another weird fish of the bottom is the tripod fish which supports itself on two long ventral fins and an extended tail lobe. It is considered a predator and sits facing the current with elongated ray fins directed ahead, possibly to detect an approaching victim. Although there are some monsters in the deep sea like the giant squid, most life is very small, possibly because food is so scarce. In the absence of light, the more traditional methods of predation, like chasing, are useless. Feeding is more dependent on trickery and chance encounters. As a result the strong bones and firm musculature of the rapidly swimming fishes above is generally not seen.

But there are some more palatable fish, such as the sablefish, that is commercially fished near the surface in the colder waters of the north Pacific, but that can also be found at abyssal depths in lower latitudes.

Most of the life in the abyss is concentrated on or near the bottom. The soft oozes and sediments are often streaked with the trails of both burrowing creatures and epifaunal organisms that forage along the surface of the floor, such as the brittle stars which sometimes live in very large aggregations. These are relatively small animals, measuring no more than six inches across, that feed on molluscs, small crustaceans, polychaetes, other echinoderms, and detritus.

There are also sea urchins, heart urchins, sand dollars, and deep-sea cucumbers, the

Deep living. Many creatures live in the solitude of the deep sea, including the rockfish (opposite, top), the hatchetfish (opposite, center), the squid (opposite, bottom), and gorgonians (right).

holothurians. With their tubelike bodies, some of these holothurians gather in organic matter, living or dead, by means of tentacles near the mouth end. Other sea cucumbers take in sedimentary material, letting their digestive systems extract the organic material. In either case, the holothurians are responsible for stirring the sediment.

There are several kinds of worms living in the abyss, such as the acorn worm, which may be the creature responsible for some of the intricate spiral patterns that have been photographed on the sea floor.

Chapter IX. The Hadal Zone

The very deepest parts of the ocean, which have been viewed only a very few times through the windows of bathyscaphs, are called the hadal zone. The total darkness, the intense pressure, the constant cold conjures images of a Hades quite different from the fire and brimstone of the Hell of fundamentalist preachers. But the fire and brimstone are there, provided by the earth itself when molten rock bursts through the thin crust and spews forth with a grisly glory.

The volcanoes are surface indications of the geophysical activity deep in the bowels of the planet. Volcanoes, earthquakes, perhaps even the deep-sea trenches are the tools with which the earth is remodeled, and its face changed. Rather than viewed as forces of destruction, which in terms of human life they often are, they may be considered as forces of building and change.

Some theorists propose that the trenches are carved along the edges of large plates which slip and slide around the globe. They say the earth is not like an onion with layer upon layer wrapped uniformly around a central core. Rather, the layer beneath the earth's crust is made up of a series of movable plates and the deep-sea trenches are the places

---

"Trenches may represent regions where one crustal plate is forced below another, dragging the bottom downward."

---

where two plates meet, one descending under the other and thus creating the mammoth depressions that can extend 36,000 feet below the level of the sea. It would seem logical that mountain building, volcanoes, and earthquakes would occur at the edges of these plates; instability would coincide with areas of collision between two solid objects. Graphically illustrated, the slope from Aconcagua, the highest peak in the Andes, down to the bottom of the Peru-Chile Trench represents a vertical drop of more than 47,000 feet across a horizontal distance of less than 100 miles. If the plate theory holds, then the westward movement of the plate with the South American continent on its back has left little opportunity for a broad, gently sloping continental shelf to exist.

With all the activity and instability of the hadal zone, life persists in the area, although at a seemingly much reduced level. The term biomass is often used to describe the amount of life in an area. Biomass is defined as protoplasm weight or the amount of living organisms in one area, usually expressed in terms of grams per square meter of ocean bottom. In the intertidal zone, for example, the biomass can range anywhere from 100 to 5000 grams per square meter, while at a depth of 12,000 feet, the biomass is only about 5 grams per square meter. And at the bottom of the Tonga Trench, 30,000 feet below sea level, the biomass is one one-thousandth of a gram per square meter.

This does not mean that trenches are deserts. In fact, it may be only that it has not yet been possible to accurately measure the amount of life that is there.

The shapes of creatures living in the trenches are probably not very different from those to be found in other parts of the deep ocean. The trenches probably do not form a separate environmental niche as do coral reefs, estuaries, and neritic surface waters.

Life on the floor. Despite the low average biomass figure, there is life in the hadal zone, as this sablefish attests. There are also starfish and the tracks and trails of other types of benthic organisms.

Vulcan's Ring of Fire

Thin wisps of blue smoke dance toward the sky to the accompaniment of groans sounding like underground breakers. Plumes of yellow molten rock begin spewing upward like a fountain of fire reaching 50 feet into the air, then 100 feet, perhaps 2000 feet. Black smoke belches from the guts of the earth—a snowfall of volcanic ash.

The scene could be any one of a hundred places along the edge of the Pacific Ocean, the area called the Ring of Fire. It could be Krakatoa in Indonesia, or somewhere in the Aleutian Islands, or perhaps a rejuvenated South American volcano. It could have been ages ago before there was anyone there to see it; or centuries ago when there were plenty of people to be killed by it; or a few years ago when a photographer was there to capture the event on film.

The presence of volcanoes, often on island arcs and usually occurring in association with deep-sea trenches and earthquake activity, is indicative of the earth's restlessness far below the surface. There molten basalt rocks flow around the globe, pushing up in many places at once and perhaps forming several new seamounts, or erupting in one place in most spectacular fashion.

The Ring of Fire girds the Pacific, hopping from island arc to continent to isolated peak. It is not a continuous circle, for there are wide gaps along the coast of North America.

Volcanic activity is not confined to the Pacific, nor even to ocean areas for that matter. Land volcanoes spew rocks containing some granite particles, while this material is usually absent from ocean eruptions.

The North Atlantic has been an area of remarkable vulcanism in recent times. In 1963 Surtsey Island was created as it rose out of the sea near Iceland. Ten years later erupting volcanoes spewed forth over Iceland, indicating considerable geophysical activity beneath the floor of the Atlantic.

Kealakomo, Hawaii. *Lava flows into the Pacific Ocean (left) following an eruption in 1971. The Hawaiian Islands are of volcanic origin but are not part of the Ring of Fire.*

Getting the story. *Eruptions (opposite page) have given rise to many folktales about fire gods, but now these spectacular occurrences help scientists in their study of the earth's history.*

The Mystery of the Trenches

Earthquakes, volcanoes, and tsunamis are some of the most destructive forces on earth. For some their origins lie in the hadal zone of the ocean, whose very name is derived from Hades, the Greek word for underground but our word for Hell.

Earthquakes, volcanoes, and tsunamis often occur in areas of the ocean where deep-sea trenches are found. The deepest of these trenches is the Marianas Trench in the northwest Pacific, where Jacques Piccard and Don Walsh reached the depth of 35,800 feet in the *Trieste* in 1960. One of the first things they saw at that depth was a flounder-like fish, and a shrimp, which helped soften the hadal connotations of the trenches. There are several of these large depressions, much longer than they are wide, in the Pacific, which have depths greater than 30,000 feet. They are always adjacent to a volcanic arc. There are trenches in the Atlantic, such as the Puerto Rico Trench and the Romanche Trench, but these are much shallower. The trenches tend to have steep walls with a narrow floor, although those near land masses have wider floors, probably because sediment has filled in the bottom. The slope of the V-shaped wall usually averages about 8° to 16°, getting steeper near the bottom. But rather than a continuous slope, the walls

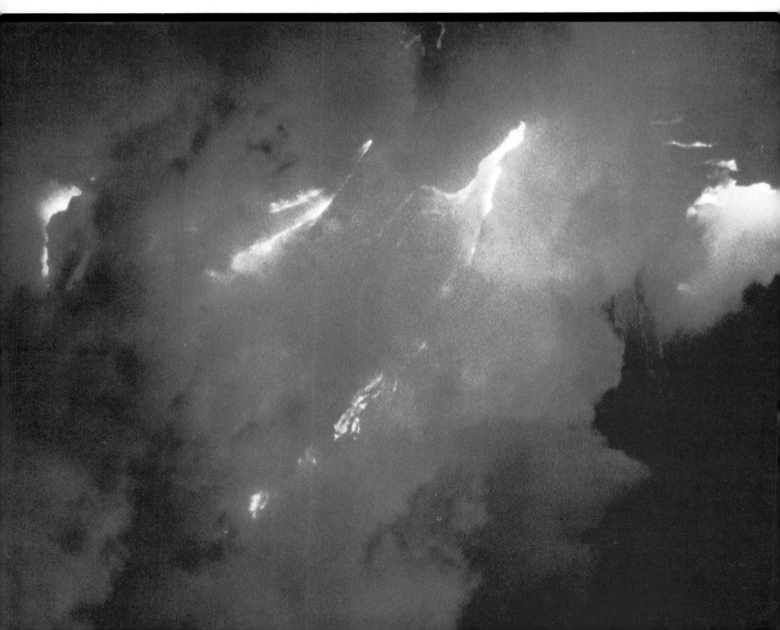

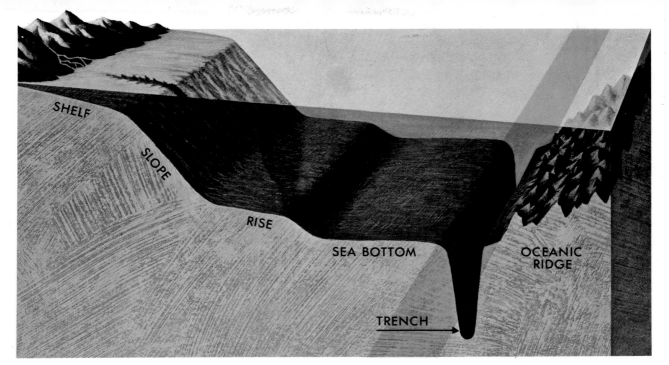

Profile of the hadal zone. The deepest holes on earth are the ocean trenches, which plunge into the earth more than 35,000 feet, deeper than the highest mountains are tall.

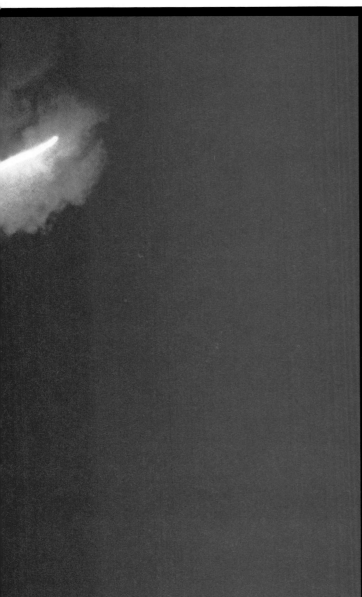

Lava flow (left). Volcanoes in the Pacific are thought to have some inherent connection with trenches.

are made of huge steps with many vertical walls. The Tonga Trench in the southwest Pacific, which is more than 35,000 feet deep and 33 miles wide, is thought to have walls that slope at a 45° angle.

In the Pacific, especially, mountain ranges or ridges are usually found on the island arcs and continents flanking the trenches. It is on these island arcs and on continental coasts that many of the world's active volcanoes are found. Earthquake activity is associated with trenches, which some believe is evidence of intense horizontal movement beneath the crust; such activity is assumed to be the force which touches off the destructive waves called tsunamis.

The trenches, at the edges of volcanic and mountain-building activity, separate the stable seabeds from areas of continental activity, as though they were moats separating the calm from the wild.

113

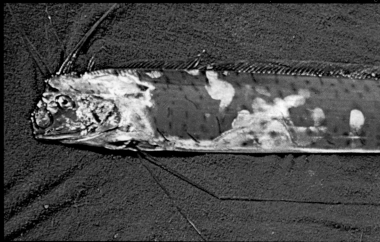

Up from the bottom. *These eellike creatures of the abyss display the long-pointed tails that are common to many fish living at great depths.*

Decked deep-sea fish. *The oarfish got its name because of its long thin shape. Its body may be eight feet long and only two inches wide.*

Outposts of Life

The hadal zone is not lifeless, although there certainly seems to be far fewer species and individuals in the trenches than there are in other parts of the sea.

Among the adaptations most deep-sea creatures have made to abyssal life is the reduced use of eyesight. Most of the fish of the bathypelagic region do have eyes, often enormous eyes, indicating there are enough flashes of bioluminescent light to keep the eyes from degenerating totally, as has occurred in some terrestrial cave dwellers. There is one bristlemouth species, *Cyclothone obscura,* which is believed to be without light-sensing organs. In the anglerfish group there is a species in which the male as a juvenile has very large eyes that later degenerate. It is thought that this is a result of its changing life-style. The large eyes may be used in locating a female, but are not needed after the mate is located and the male becomes a parasite on it. The females do not possess eyes that are as well developed as those of juvenile males.

In other fish, such as gulper eels and black bristlemouths, both sexes have small eyes, while the species only sometimes found in deep seas, like brotulids, have much better developed eyes. The bathypelagic fish which hover near the bottom—they possess no swim bladder—often have lost all of their eyes except the retinae.

Many of the animals which live on or near the bottom have reddish or pinkish coloration. This may be because red light waves are absorbed in the uppermost layers of the ocean, and as a result red-tinted creatures would appear no different from black creatures because no red light would be available to be reflected.

Of the deep sea invertebrates, there seem to be two major life-styles. Some strain or grab food from the currents and, as a result, most often possess delicate extensions of their bodies. These include the vase-shaped glass sponges that look like antennae with symmetrical branches on a main stalk; sea pens on elongated stalks that branch into a star-shaped structure at the top; and crinoids

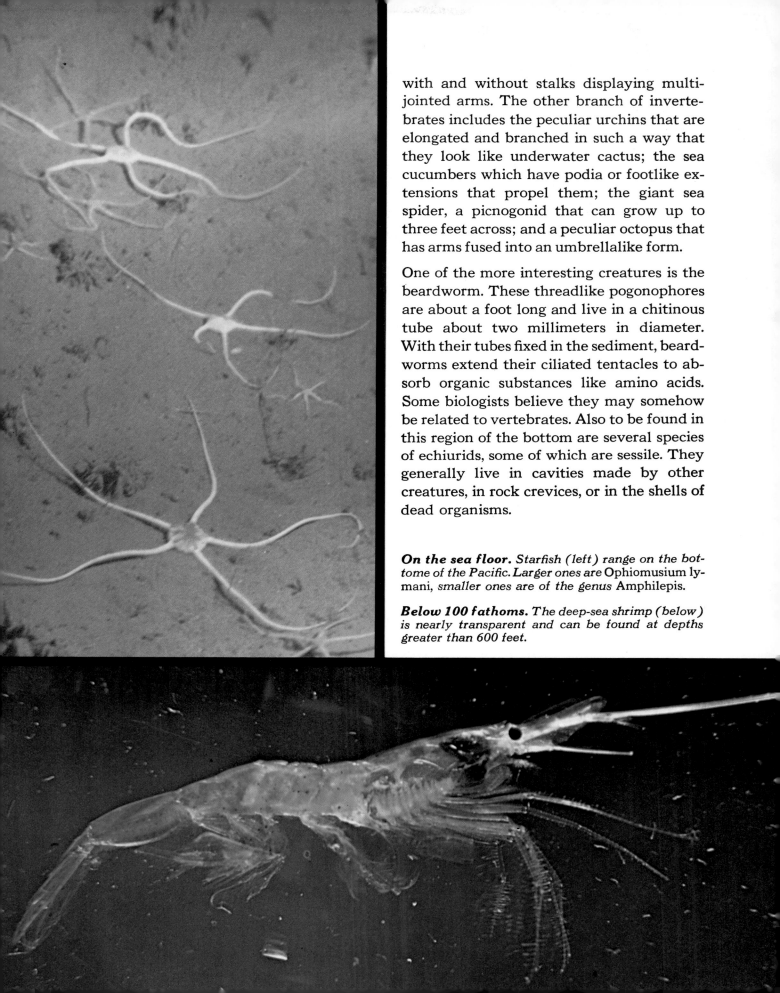

with and without stalks displaying multi-jointed arms. The other branch of invertebrates includes the peculiar urchins that are elongated and branched in such a way that they look like underwater cactus; the sea cucumbers which have podia or footlike extensions that propel them; the giant sea spider, a picnogonid that can grow up to three feet across; and a peculiar octopus that has arms fused into an umbrellalike form.

One of the more interesting creatures is the beardworm. These threadlike pogonophores are about a foot long and live in a chitinous tube about two millimeters in diameter. With their tubes fixed in the sediment, beardworms extend their ciliated tentacles to absorb organic substances like amino acids. Some biologists believe they may somehow be related to vertebrates. Also to be found in this region of the bottom are several species of echiurids, some of which are sessile. They generally live in cavities made by other creatures, in rock crevices, or in the shells of dead organisms.

On the sea floor. *Starfish (left) range on the bottome of the Pacific. Larger ones are* Ophiomusium lymani, *smaller ones are of the genus* Amphilepis.

Below 100 fathoms. *The deep-sea shrimp (below) is nearly transparent and can be found at depths greater than 600 feet.*

Chapter X. Province of the Unknown

To understand the origins of the oceans, it is necessary to understand the formation of the earth itself. The molten core, the deep layer of dense rock known as the mantle, and the relatively thin crust all hold evidence of the sequence of events which is an ongoing process continuing to this day.

The crust, because it is readily available for study, has yielded much to exploration but not nearly as much as there is to know. And in reality the crust is just a very thin skin on the face of the globe. The crust has two distinct components—granite and related rocks, which make up the continental crust,

---

"The oldest islands in the Atlantic are only 120 million years old, which would put their formation in the Cretaceous period."

---

and basaltic rocks, which are found under the oceans and beneath the continental crusts. The chemical composition of these rocks, their origin, and the manner in which they came to rest in their present position all help reveal forces of earth building.

Geologists have developed a time scale of the earth's history based on data obtained from different rock formations. The clues may be fossil plants and animals in different stages of evolutionary development, minerals which lose their radioactive properties at known rates, or ancient magnetic orientations. The scale begins with the pre-Cambrian deposits, those rocks which are 600 million years and older. The borderlines between the eras are indistinct, and since not all rocks are found in all places, the record has to be put together piecemeal. But the surprising thing is that although there are rocks on the continents 600 million years old, there is no evidence to show that the

oceans, in the shape they are in today, are anywhere near that old. The oldest islands in the Atlantic are only 120 million years old, which would put their formation in the Cretaceous period. However, when the *Glomar Challenger* began its drilling explorations in connection with the Joint Oceanographic Institutions for Deep Earth Samples (JOIDES) project in 1968, it recovered the oldest rocks ever taken from an ocean basin. These Atlantic sediments were 140 million years old, from the Upper Jurassic period. Since these were sediments, they must have been laid down in an ocean bed that was somewhat older, but there is no evidence that the Atlantic basin is anywhere near as old as the rest of the earth.

The sea, as has so often been said, is restless. The endless motion of waves, deep water circulation, and surface currents are only part of the story, for the very bed in which the ocean lies is not the immovable object that rock has always been assumed to be. The probability is that the sea floor has moved considerably, if not constantly, since the basins were formed. The assumption is that the floor of the Atlantic spreads from the midocean ridge toward the continents flanking it on either side. This motion, apparently being fed by material from beneath the crust, would help account for the lack of older sediments and the relative youth of the basins themselves, if they are going to be dated solely on the age of the rocks forming the basins. There are more questions than answers, and one of the provinces of the sea has to be the Province of the Unknown.

The Western Hemisphere. America is distinct from the Old World and may be becoming more so, since scientists have suggested that the Atlantic Ocean is continuing to spread from the middle.

A Trip inside the Earth

Starting at the floor of the ocean, if we could proceed directly to the center of the earth, we would pass through several layers composed of a variety of materials. These layers do not have the same thickness everywhere, but the order is uniform. First, there is a layer of unconsolidated sediment, perhaps 3000 feet thick, then the so-called second layer of consolidated sediment approximately twice as thick. After that follows a layer of basaltic rock, then the much denser mantle of the earth before reaching the core. The core is sometimes divided into the outer core and the inner core.

If that same trip were taken from a start on land, there would be—after the thin layer of surface rocks and soil—a thick, thick layer of continental crust made up primarily of granite and related rocks before encountering the basaltic layer, the same one that underlies the sea. Then come the mantle, outer core, and inner core in that order.

In other terms, the upper edge of the earth's mantle is much further below the continents than it is below the bottom of the sea— about 19 miles further below.

The edge of the mantle is fairly well determined, since earthquake shock waves speed up considerably when they reach the mantle. This alteration in the wave pattern is called the Mohorovicic Discontinuity, after the Yugoslavian geophysicist who discovered it. The Moho, as it is called, is the frontier of the mantle.

In terms of a global scale, the distance from the surface of the earth to the Moho is only a thin line compared to the thick segments that constitute the mantle and central core. But the existence of different thicknesses and types of crust beneath the oceans and the continents helps determine a couple of things. First, there is little evidence for lost continents having been swallowed up by a gaping sea floor. And second, the relatively thin layers of sediment on the ocean floor indicate that the oceans' basins have somehow been cleaned of sediment or else the basins themselves are not as old as the entire earth is.

The core is a very puzzling part of the earth, since it obviously is not readily accessible to scientists for observation. But because of the earth's equatorial bulging and flattening at the poles, it is apparent that most of the earth's mass is near its center. The great pressure and intense heat near the earth's core is the result of the heavier material there, which probably trickled down when the infant earth was still in its molten state. Because of the high temperature at the core, the outer portions of it are most likely molten, but at the very center, the core is still solid because the tremendous pressure prevents the material from melting.

Concentration. The earth is really made up of a series of concentric layers composed of different materials. The core is very hot and very heavy, while the mantle is solid rock, though it has a plasticity to it and is much cooler and much less dense than the core. The crust is the thinnest layer, containing the lightest rocks.

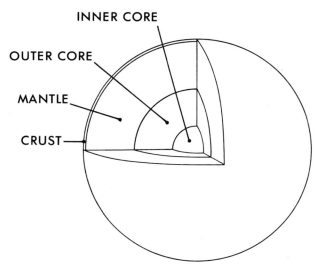

INNER CORE

OUTER CORE

MANTLE

CRUST

Of Sima and Sial

Before man reached the moon and was able to sample and study its rocks, there were many fanciful theories about it; one of the most popular suggested that it had once been part of the earth. This theory, its supporters said, explained the presence of the Pacific basin, with its vast stretches devoid of landmasses. But the rock samples obtained from the moon proved to be of light material, the stuff of which continental crusts are made, and not the heavy basalt that is found at the base of oceans. So the notion that the moon was formed in the Pacific basin and then was pulled from the earth is now destroyed forever.

The two predominant types of materials in the earth are mixtures called sial and sima, and their different characteristics help determine the domain of the sea and the realm of the land. Sial, a term derived from the chemical symbols for silica and aluminum, its major constituents, is light-colored, with a density of 2.7, and composed of granite and similar rocks. Sima, made up primarily of silica and magnesium, has a specific gravity ranging from 2.9 to 3.4. Dark-colored basalt typifies sima. The continental crust consists of sial resting on sima, while beneath the oceans there is only sima, no sial.

The different densities of the rocks, taken in consideration with the thickness of the sial layer beneath the continents, confirms the principle of isostasy, according to which the landmasses are buoyed above sea level because they float on heavier material.

Into each layer, however, a little rock must intrude, and there are patches of sial rock on the floor of the Atlantic and Indian oceans, like the Seychelles Islands, and perhaps in the Pacific around Easter Island. There are numerous outcrops of basaltic rock in continental regions, but one of the more spectacular is on the Mediterranean island of Cyprus. The island, which rose from the sea two million years ago, displays not only basaltic rocks, but layers of what many believe to be material from the mantle itself, from below the Mohorovicic Discontinuity.

Inside the earth's skin. The earth's crust (below) is made up of two distinct parts, the continental rocks, or sial, and the denser basaltic rock, or sima, which coats the mantle and underlies both the continents and oceans. There is little or no sial in the oceans, except for the sediments which have been eroded from the continents. These are very thin layers which become compressed and compacted over hundreds of thousands of years, only to be thrust up again (opposite page) by mountain building forces inside the earth which can twist, turn and fold the rocks in contorted patterns. Once they are out of the water, they will again be subjected to erosion and the cycle will start over again.

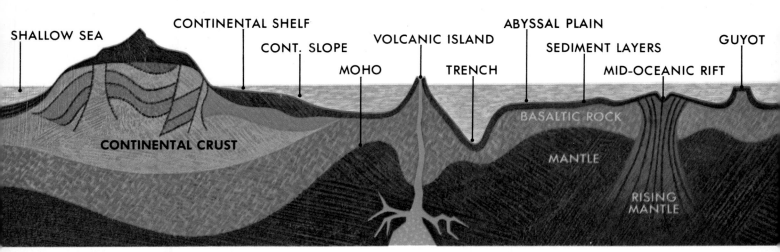

SHALLOW SEA   CONTINENTAL SHELF   CONT. SLOPE   VOLCANIC ISLAND   MOHO   TRENCH   ABYSSAL PLAIN   SEDIMENT LAYERS   MID-OCEANIC RIFT   GUYOT   CONTINENTAL CRUST   BASALTIC ROCK   MANTLE   RISING MANTLE

Boring into the Earth

The history of oceanography goes back to the time when man first speculated on the nature and origin of the oceans. Through the ages the ancients measured, guessed, proposed, and theorized as best they could. The Phoenicians, Portuguese, Chinese, or British, and all the people before, after, and in between them were able to add only as much knowledge as their tools would allow. Even with modern technology we are still using microtools to study the magnaworld of the oceans. But technology did allow modern man to begin positive speculations about what may lie beneath the oceans.

The intriguing way in which earthquake shock waves speed up once they pass the

Underwater research. The 400-foot deep-sea drilling ship Glomar Challenger (*right*) *is used in a project to bore into the rocks beneath the ocean. The main tool on the ship is the drill itself (*above*) suspended from a 150-foot derrick. Some of the results have been the oldest rocks ever recovered from the Atlantic, the deepest hole ever drilled into the ocean floor, and the finding of the first oil deposits located in open water away from coastal areas.*

Mohorovicic Discontinuity prompted scientists to propose drilling a hole—the Mohole—into the mantle and extract a core sample. Since the Moho is so much closer to the earth's surface beneath the oceans, the decision was to go through water, and then through the relatively thin basaltic oceanic crust in order to reach the mantle. This seemed to make more sense than drilling through all the granite rock which underlies the continents. The project provided for a $100 million budget and was abandoned before it could yield much useful information. But the idea of the Mohole spawned interest in the less ambitious but more practical plan to drill into the upper layers of the earth's crust beneath the oceans. The ultimate purpose of the Joint Oceanographic Institutions for Deep Earth Samples (JOIDES) project was to determine the age of ocean basins and the processes by which they developed.

The problems in drilling into the earth's crust include such practical matters as how to keep the ship from drifting in gale-force winds at the surface when the drill is 20,000 feet below the surface penetrating rock 2000 feet thick. But thanks to "dynamic anchoring" the *Glomar Challenger* has been able to recover the oldest sediments ever taken from the ocean basins, dating back 140 million years to the Upper Jurassic period. In subsequent drilling, it brought back samples of basaltic rock which was younger than the Upper Jurassic sediments, but which indicated that the basement rock of the ocean basins had been reached.

Building an Ocean

The shape of the shoreline is not static, the level of the sea is ever changing, and the configurations of the ocean basins are unstable. The oceans, as we know them, have not been around for all time. It is very likely that there was no Atlantic Ocean before 150 million years ago, until after South America and Africa began to grow apart. The mechanism for the separation of landmasses and formation of ocean basins could be continental drift, sea-floor spreading, New Global Tectonics—any of these, all of these, or some idea as yet unstated.

The fact remains that there are no islands older than 120 million years in the Atlantic. And in other parts of the world new oceans may be forming. The Arabian peninsula was joined to the African mainland about 20 million years ago, but since then the Red Sea has been growing. In North America, some theorists say, California will sink into the Pacific or at best become an offshore island, the Gulf of California expanding as land slips along the San Andreas fault.

If the Atlantic Ocean basin has been created from a rift emanating in the midocean ridge system, the depression was then swiftly filled in by the onrushing sea back in that Jurassic time. Shortly afterward sediment had come pouring into the ocean, so that the bottom then resembled the muddy floor of the Red Sea as we know it. And as the Atlantic grew wider, the edges of the newly created continents must have been pushed toward the sky, as are the African shores bordering the Red Sea. And as the ocean bed gets wider, it begins to sink, creating a deep ocean, but one with shallow shore areas as the erosion begins to take its toll on the upthrust continental coasts.

The idea of sea-floor spreading does not necessarily mean that the continents were being pushed apart by its force, for it is more likely that the sea floor spreading from its middle in two directions folded under the continents as it reached their borders. The evidence for such a sequence of events is more complete for the Atlantic than for the Pacific or Indian oceans.

Rift valley. Two views of the San Andreas fault are shown at Drake's Bay (right) northwest of San Francisco and in the Santa Cruz Mountains (below) south of San Francisco, where the fault itself is visible as a thin colored line in the middle of the picture.

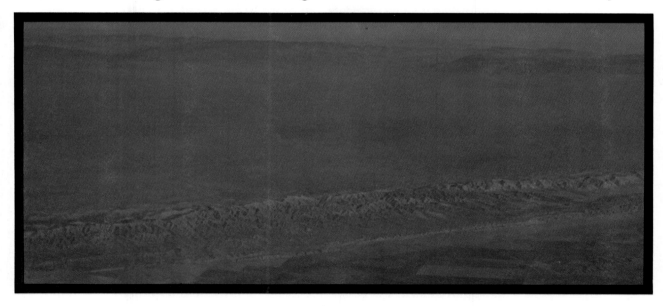

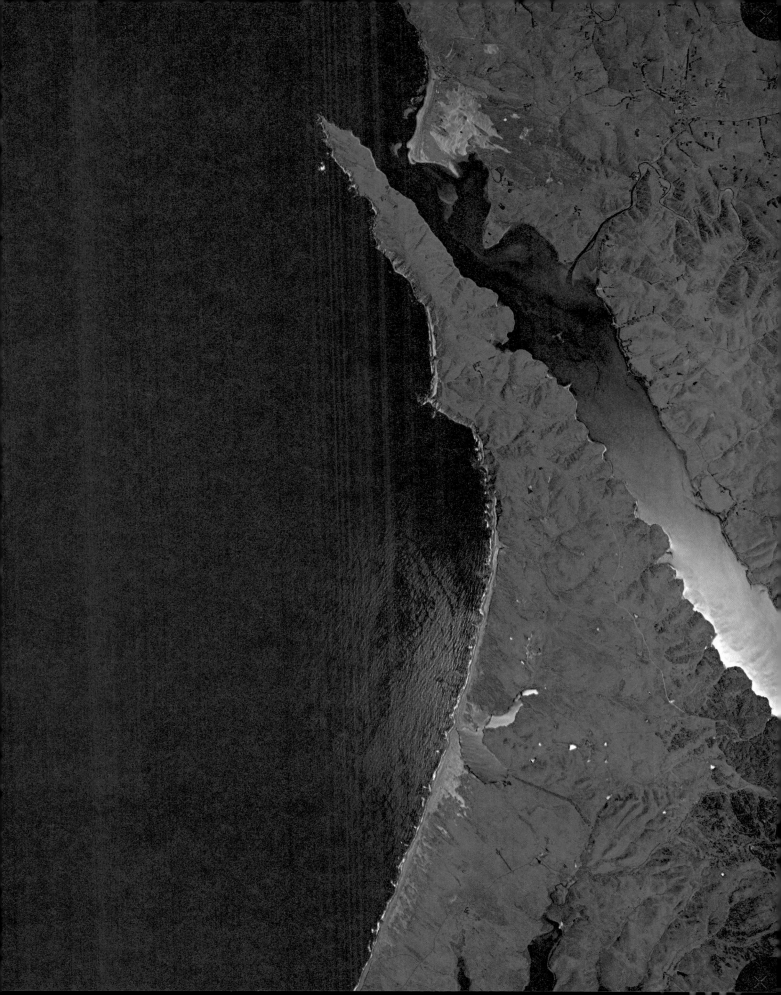

Chapter XI. Shaping the Earth

When a child asks, "Where did I come from?" the answer most often deals not with "where" but "how." The same is true of the earth. Questions of where the continents or oceans came from are usually answered with the hows of continent building or ocean forming. The difference is like that between knowledge and speculation.

Since the very symbolic and somewhat misleading explanation found in Genesis, for most of history men have accepted the idea that the earth has been as it is today since the time of its creation.

Then, as the age of science dawned, in the seventeenth century and perhaps before, man began thinking about the possibility that the continents were not always as isolated as they seemed to be. With so much change in the world, why should only the continents themselves be stable?

The preliminary work and reasoning took hold early in this century, especially in the thoughts of Penck and Wegener, two German geologists, the most prominent of the early "supercontinent" supporters. They proposed that all the continents were once

---

"The moving forces behind, or rather under these plates, are convection cells formed by heat rising from the earth's interior through the mantle."

---

part of a single landmass called Pangaea. The idea did not die with them, for many others, serious thinkers and charlatans alike, have built similar models of the earth as it might have been early in geologic time.

The hypothesis was supported by evidence, not only of sedimentary beds, but also of fitting shorelines, such as those of South America and Africa. There were also indications in fossils, glacial records, paleomagnetic rocks, and radioactive dating of organic material and inorganic rocks. The evidence was mounting. The only thing lacking was the understanding of the machinery that could cause, or allow, the continents to drift.

Plate tectonics was the answer. The idea is that the earth's crust is composed of a series of plates or blocks, 45 miles thick under the oceans and 75 to 90 miles at the continents. The boundaries are determined by areas of activity, such as earthquakes and volcanoes. When these plates meet with force, they create mountains like the Himalayas, or they fold one under the other to form deep trenches. In places where the plates separate, new material is pushed up from the center of the earth to fill the void, as is apparently happening along the midocean ridges. The moving forces behind, or rather under these plates, are convection cells formed by heat rising from the earth's interior through the mantle. As new material is added, an equal amount must be subtracted, most likely being forced back toward the interior of the earth under the continents or in a downwelling action in the areas of deep-sea trenches.

Plate tectonics, while just a theory, does provide a reasonable explanation as to how the continents could have been a single landmass, only to be broken apart as the earth continues to metamorphose. But many questions still remain without answer.

Himalayan Mountains. These mightiest of all mountains were formed, according to the theory of continental drift, when the Indian subcontinent collided with the Asian mainland sometime back in the geologic past, about 65 million years ago.

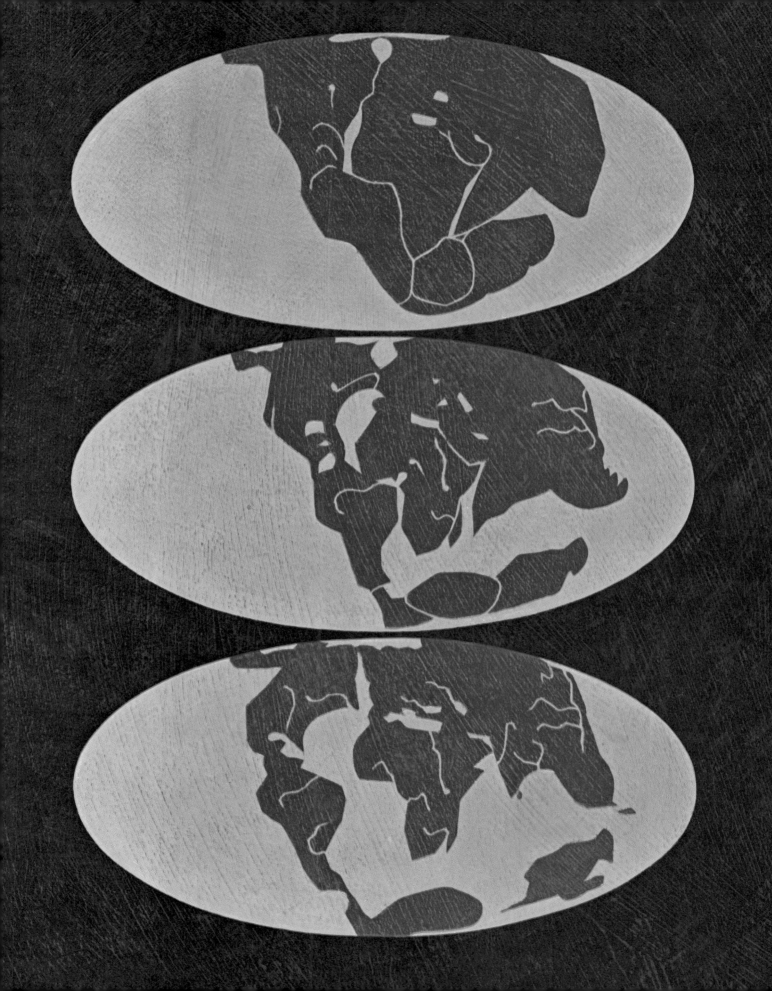

Making of a Puzzle

Solid as rock! What could be more absurd than solid granite continents floating about the face of the earth, banging into each other here, narrowly avoiding each other there, and gently joining forces to form a super landmass in still another place?

Yet sober-minded scientists very early proposed that the continents were not always in their present positions. In 1620 Francis Bacon considered the possibility, as did P. Placet 50 years later. It was another 150 years until Alexander von Humboldt revived the idea, which is only now being accepted. An American, F. B. Taylor, thought the continents had moved from polar regions toward the equator, while the German geologist Alfred Wegener envisioned the continents as "ships of sial plowing through a sea of sima." By 1937, after years of studying similarities between South American and African fossils, Alex L. Du Toit, a South American, wrote the book *Our Wandering Continents*. By 1944, A. Holmes had studied glacial climates in the Southern Hemisphere and agreed that the continents must have moved. He collected data in Argentina, Madagascar, subequatorial Africa, and southern Australia. His efforts and those of Du Toit set the tone for a difference of opinion among geologists which has lingered until the present. It is the Southern Hemisphere scientists who generally support the notion of continental drift, mainly because the evidence was so plentiful in their areas, while Northern Hemisphere geologists were skeptical, largely because of a lack of evidence in North America and Europe.

The presence of sedimentary rock in the interior of the continents intrigued these theorists and led them to the conclusion that the continents had moved. Great mountain ranges like the Rockies and the Himalayas contain beds of sedimentary rock of the very kind that is laid down in shallow coastal seas. When these sedimentary beds were correlated with other rock layers and were matched with the geologic time scale, no end of confusion resulted. It was impossible to plot the rise and fall of the sea in such a way as to explain the presence of sedimentary beds where they were known to be. Impossible, of course, only if the continents had never changed their positions and had always been where they were presently found. Thus, the stage was set for the puzzle. Globes were constructed and paper models of continents were made with beds of similar sedimentary rocks colored one way, or like glacial evidence shaded in, or fossil remains indicated by cross-hatching. The scientists then pushed these pieces of paper around and moved them every which way in an effort to correlate the markings on the different continents. Glacial data in Australia was matched to that of Madagascar. Fossil remains in southwest Africa were paired with those of eastern South America. All sorts of theories were propounded for the continents moving this way and that, or having been attached to one or another in any variety of ways. The evidence became stronger and stronger that Wegener's idea of drifting continents was only part of the answer. It seemed more and more likely that not only had the continents moved, but that they all had been part of the same ancient landmass at some time in the distant past.

Wegener's worlds. The German geologist Alfred Wegener constructed a three-stage model for the breakup of the supercontinent Pangaea. The splitting began about 300 million years ago (top) when large areas of Pangaea were covered with shallow seas. Then, after about 240 million years (middle), the continents began to take their distinctive shapes. About a million years ago, the continents were formed for all practical purposes (bottom), but had not yet drifted across the earth to the position that they occupy today.

Putting the Pieces Together

The idea that all the continents had once been joined together in a supercontinent was born when maps of the world gained accuracy; some investigating minds thought of piecing together the landmasses like parts of a global jigsaw puzzle. It was shoreline against shoreline, sedimentary bed against sedimentary bed. The theory gained cre-

dence as more and more was learned about the continental shelves, the true outlines of the continents. The pieces then began to fit a little better.

However, no matter how good the reconstruction of Pangaea, Wegener's name for the supercontinent, there were always parts which couldn't be accounted for. If, for example, South America and Africa were joined and there was no Atlantic Ocean, it

was probable that North America was joined to something, too, whether it was Europe or the northwestern coast of Africa. But either way, there was no room for Mexico and Central America in the reconstruction. And the southern coast of Asia just didn't seem to fit anywhere.

Despite such problems, supporters of continental drift saw too much evidence. There was a general feeling that similarities in out-

Fit between the shorelines. First the outlines of the continents and then the continental margins provided support for early backers of the theory of continental drift.

lines of the landmasses and the continental margins were more than just a coincidence. But whether sedimentary beds, shorelines, or continental margins were considered, the main problem was to bring the supercontinent into a reasonable shape.

Pulling the Pieces Apart

When solving a puzzle, one begins with the easiest parts, so in reconstructing the primeval continents before they drifted, South America and Africa were always joined together first. But from there on, geophysicists met one problem after another, and more hard thinking was necessary.

Progress was made when two supercontinents were proposed, Gondwanaland and Laurasia, instead of the monolithic Pangaea. These two paleocontinents broke up independently of each other and gave rise to our present landmasses. But no amount of twisting and turning these two pieces could correlate all the physical and fossil findings that were accumulating. Certainly, there was evidence that Europe and North America, for example, had less in common than did Africa and South America, at least in more recent geologic times. But in considering all the evidence from all time periods, any two-continent explanation raised as many problems as it answered.

The most likely solution, as proposed by Robert S. Dietz and John C. Holden, was that at one time there was a supercontinent, Pangaea, which existed in the Permian period, 280 million years ago. But by the end of the Triassic, 80 million years later, Pangaea had split into Laurasia (North America and Eurasia) in the north and Gondwana in the south. Africa, South America, Australia, Antarctica, and the Indian subcontinent were all part of Gondwana. As the supercontinent was breaking up into two parts, Gondwana itself was being subdivided by the rift known today as the southwest Indian Ocean Ridge. This rift separated West Gondwana (South America and Africa) from East Gondwana. During the Jurassic, 170 million years ago, South America and Africa began to split, as India was drifting away from the Antarctica-Australia complex in East Gondwana.

One of the geophysical phenomena which has helped in plotting the movement of continents is the Walvis hot spot that formed during the Jurassic period. Located about 125 miles southwest of the island of Tristan da Cunha in the South Atlantic, the Walvis hot spot has remained permanent in its location. As a result, when continents moved over this thermal center, they were distinctively marked. Since the hot spot's location is

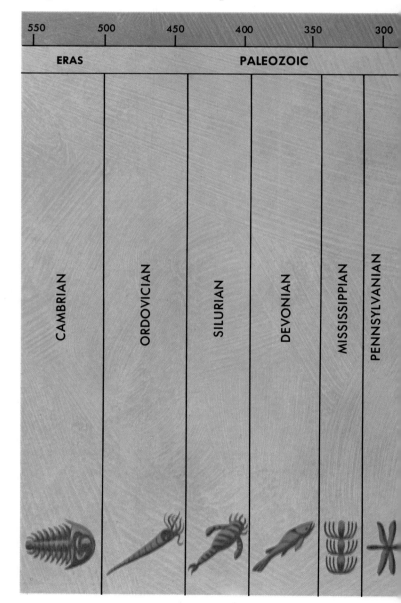

550	500	450	400	350	300

ERAS — **PALEOZOIC**

| CAMBRIAN | ORDOVICIAN | SILURIAN | DEVONIAN | MISSISSIPPIAN | PENNSYLVANIAN |

known and has remained fixed, geologists can track the speed and direction of continental movement.

By the end of the Cretaceous period, 65 million years ago, Africa and Eurasia were rotating toward the present positions, while India was heading for a collision with the Asian mainland. Australia began heading east and South America west, where it would eventually rejoin North America. Greenland, not yet an island, linked North America with Europe.

During the last 60 million years, Antarctica rotated slightly to the west; New Zealand split from Australia as it shifted its course from east to north; and the Atlantic and Arctic oceans were joined as Greenland became an island. Africa moved slightly to the north to its modern location, and India crashed into Asia, with the impact forming the Himalayan Mountains.

Reconstruction. *The supercontinent of Pangaea could have existed as recently as the Permian period, 225 million years ago, before it began to drift apart.*

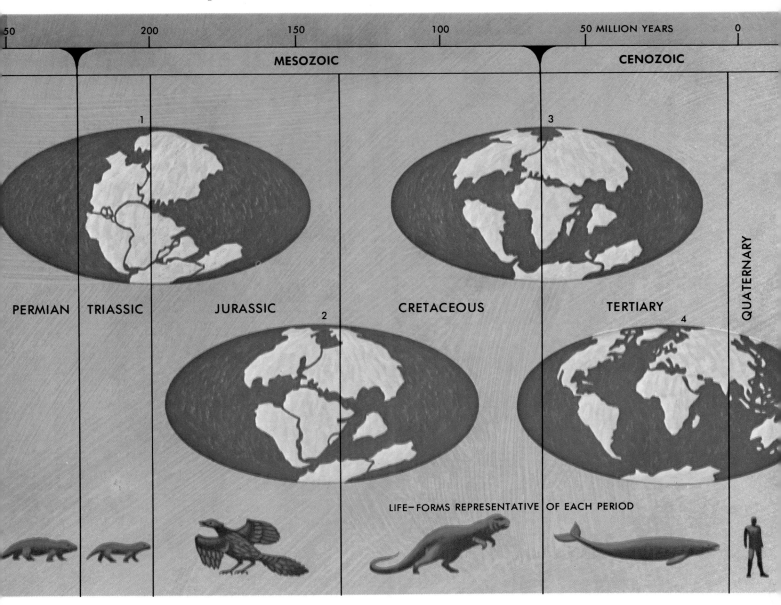

PERMIAN TRIASSIC JURASSIC CRETACEOUS TERTIARY QUATERNARY

MESOZOIC CENOZOIC

250 200 150 100 50 MILLION YEARS 0

LIFE-FORMS REPRESENTATIVE OF EACH PERIOD

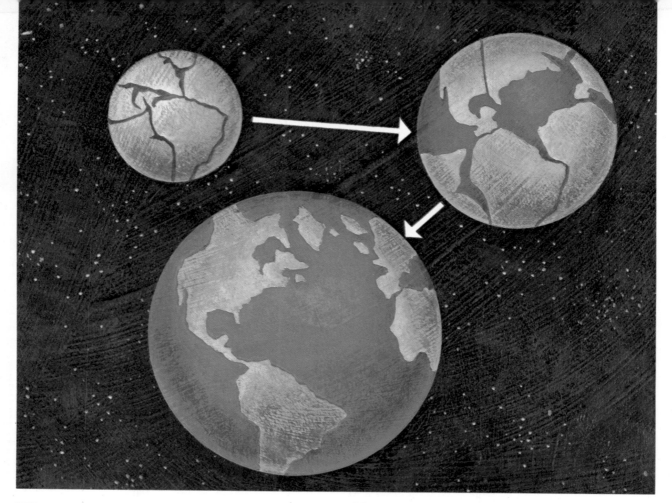

The Expanding Earth

One of the notions that has been advanced to both support and contradict continental drift is the idea that the earth itself has expanded in much the same way that a beach ball or balloon would be if it were inflated. The idea seems attractive to those who see the fit between the continental outlines but who can't accept the proposition that these landmasses have drifted around the globe independent of radial expansion.

The expanding earth supporters say, rather, that all the continents had at one time been part of a supercontinent, or perhaps covered the surface of a much smaller earth. Then, as the world expanded, the land rifted and cracked, with the basins being filled with water. The stratigraphic record on continents can be used in support.

Blow up. One explanation for the rifts and cracks in the earth's crust is that our globe was at one time much smaller and has expanded like a balloon.

The force behind the expansion may possibly be radioactive decay in the earth's interior, which would produce heat and cause expansion. On a theoretical basis, this could explain an expansion of about 60 miles in the earth's radius. Another suggestion is that the earth, held together by gravity caused during the "big bang" and subsequent expansion which created the universe, is slowly losing its gravity and the surface is expanding away from the core.

The problem comes, however, in trying to correlate these mechanisms with the fact that if the earth has expanded, it would have to be on a scale of more than 1200 miles to justify all the evidence, and none of the methods seem quite that grand.

134

Collecting the Evidence

Magnetic minerals in lava will orient themselves to the earth's magnetic field when they reach the surface. As these rocks solidify, they record the position of the rock, the direction in which it is lying, and the location of the magnetic poles. Sometimes the beds of magnetic rock are warped and folded by geologic processes such as mountain building, and paleomagnetism enables the processes to be traced by measuring the amount of magnetic deviation.

There have been occasions in the history of the earth when the magnetic poles have reversed themselves. Why, or how, this occurs no one can say for sure, but these anomalies have been well documented and, in fact, are used in support of the theory of sea-floor spreading. According to this theory, the mid-ocean ridges are in fact rifts—openings in the crust where new material surges upward. As this material reaches the surface of the crust and the water, it pushes the older rock away from the ridge. Being volcanic (and magnetic), when this molten rock solidifies, it records the earth's magnetic orientation at the time of the extrusion. If, according to these hypothesis, the sea floor is spreading, there should be regular variations in the magnetic orientations, with the oldest rock being furthest away from the ridge.

In 1963 Fred J. Vine and Drummond Matthews published a study that showed just this. They demonstrated that successive stripes, or bars, occurred on the floor of the ocean, which showed alternating patterns of magnetism, correlating with the known reversals of the earth's magnetic poles. These stripes were also correlated with the sequence of geologic ages, and it was shown that the oldest rocks were, indeed, furthest from the mid ocean ridge.

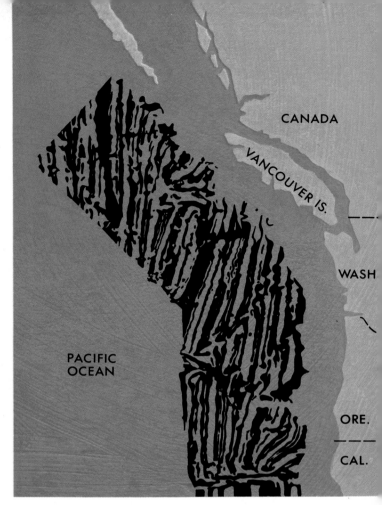

Magnetic anomalies. *A section of the earth's crust off the northwest coast of America shows the alternating stripes associated with pole reversal.*

Subsequent studies revealed that the magnetic picture was remarkably symmetrical on each side of the ridges, confirming that the sea floor was spreading toward the continents from its center. If the sea floor spreads, then continents can be moved—and such a mechanism exists on a global scale. Another consequence of the occasional alternation of the magnetic poles is that each change brings about a period when the magnetic field of the earth is negligible; this probably suppresses the Van Allen belt which normally shields us from dangerous radiation from outer space. Several times in the past creatures have been exposed to excessive radiation, which may be connected with events in the evolution of life.

135

Building a Case

Additional proof was collected as early as the 1850s when Antonio Snider began noticing the similarities in America and Europe of fossil plants dating back to Carboniferous times, 300 million years ago. Snider's work, no doubt, provided Wegener with some material for his theories. The glacial erosion and transport mechanisms brought about more evidence that Du Toit was first to interpret: beds of tillite (a consolidated glacial rubble) were found in parts of South America, Africa, Madagascar, India, Australia, and Antarctica (the constituent areas of Gondwana). These layers, ranging from the Devonian period to the Triassic, are called the Gondwana succession.

Two plant fossils in particular provide strong evidence for the notion that the various continents were once joined and later traveled. These are the fern genera *Glossopteris* and *Gangamopteris,* which have been found throughout the Gondwana succession layers, covering a period of about 200 million years. Ordinarily it takes little time and not much isolation by water for plants to diversify, and it is hardly likely that such similar ferns would develop independently over such a large geographic range.

Other evidence was provided by geochronologists who, by using radioactive-dating techniques, were able to differentiate between two beds of rocks, one going back 600 million years and the other 2 billion years. In Africa there is a sharp break between these two beds, along a line between Ghana and the Ivory Coast to the west and Nigeria and Dahomey to the east. The same age of rock beds and same sharp dividing line were found in northeastern Brazil.

Evidence. Fernlike fossils in rocks found in Africa and South America support the drift theory.

Providing the Mechanism

When the evidence became strong enough that the continents had been split and had widely changed their respective positions, the question that remained was: how?

F. B. Taylor, who was one of the early proponents of continental drift, explained that the displacement of continents may have been caused by catastrophic tidal action, perhaps when the moon first came within the earth's gravitational attraction. Another erroneous suggestion was a sudden disruption of the earth's crust when the moon was ripped away and the Pacific Ocean basin was formed. Still another outer space theory had the continents moving in response to an extraordinary gravitational pull when the planet Venus passed very close to the earth. Alfred Wegener, though, felt the mechanism lay somewhere inside the globe.

We have seen that the sea floor spreads in the following way: new material is pushed outward from the midocean ridges, and older parts of the sea floor are wedged under the edges of the continental shelf, as though the sea floor was a conveyor belt, pushing up in the middle of the ocean and heading down at the continental margins. The power plant, imagined in the 1930s by the Dutch geophysicist F. A. Vening Meinesz, could be thermal convection. According to this theory, heat from the molten, and probably radioactive, core of the earth rises through the mantle toward the crust. As a rule, heat is transferred in three ways: conduction, radiation, and convection. Convection currents are established when a fluid is heated from below, like water in a pan on a stove. The differences in density between hot fluid and cold fluid produces vertical motion deflected by the shape of the pan. Because the vertical motion is spatially patterned, rather than occurring at random, convection cells are produced as the heated fluid expands, rises, cools, contracts, sinks, and begins the cycle all over again. Even though the theory of convection cells and convection currents is based on fluids, it can be applied to the solid rock in the earth's mantle. This is because the rocks deep inside the earth have a plastic flow and behave more like fluids than solid objects. The nature of plastics is that they are brittle and shatter when subjected to sudden pressure, but flow smoothly when pressure is applied more gently.

The movement of convection cells, as they rise in the mantle toward the crust and then move parallel to the edge of the crust before descending, could drag and move the lighter

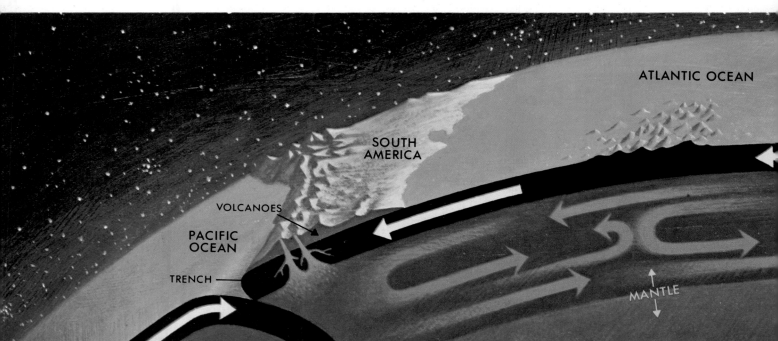

ATLANTIC OCEAN

SOUTH AMERICA

VOLCANOES

PACIFIC OCEAN

TRENCH

MANTLE

parts of the crust. As a result, the continents would be located over the areas where the convection cells begin their descent toward the core of the earth.

The ocean ridges then are areas of upwelling, corresponding to the rising of convection cells in the mantle. The outer edges of the continents, either slopes or deep-sea trenches, or both, are areas of downwelling, where the convection cells are descending. The volcanic and earthquake activity along the meridian ridges tends to confirm this hypothesis, as does the age of the volcanic rock on the floor of the ocean.

Thermal convection not only provides the mechanism for continental drift, but it helps explain why the irregular shape of the earth's surface is maintained in the face of what would be a natural tendency to flatten out.

Many geologists believe convection currents play a major role in mountain building. This geosyncline theory holds that mountain ranges start as troughs or synclines, located in areas where convection cells downwell. After the synclines have filled with sediment, tension built up by action of the current causes the earth's crust to crack and rift in the area of upwelling, and to push up the area that had once been a trough, folding and warping the sedimentary rock layers.

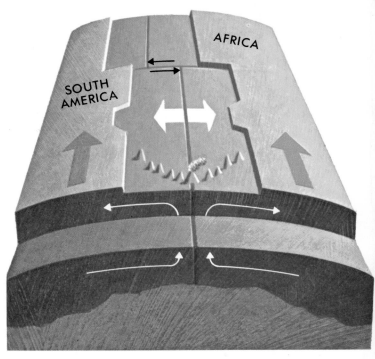

Building blocks. *Two schematically drawn continental plates (above), representing South America and Africa, illustrate how movement of the earth's crust might occur and might be recorded. As new material from the center of the earth is pushed up in the middle of the Atlantic Ocean, the continents are pushed apart. But they also move in other directions, perhaps as a result of the rotation of the earth, and the result are normal geologic faults such as transform faults. Plate movement is recorded as the blocks pass over a hot spot in the mantle, which forms volcanic mountains in the plates. The hot spot, unlike the plates, has not moved and has formed several volcanoes as the plates moved over it. The mechanism for all this motion is thought to be convection cells (below) established by heat rising from the core of the earth through the mantle.*

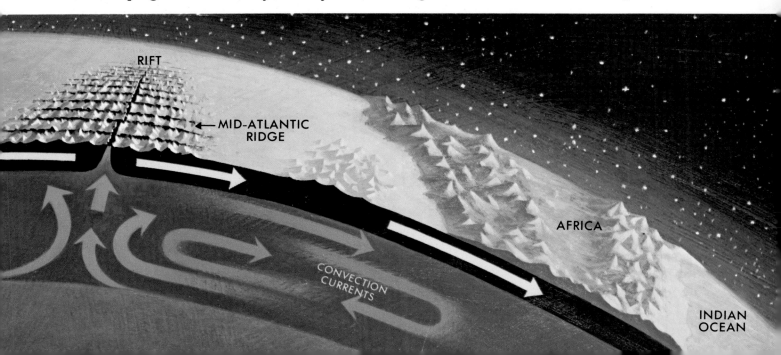

Getting It All Together

The surface of the earth is apparently neither expanding nor contracting, only altering its shape. However, there has to be an explanation for the existence of belts of earthquake activity and vulcanism along the midocean ridges, deep-ocean trenches, and certain continental areas.

One possibility, if the earth's surface is staying the same size, is provided by plate tectonics, which governs the building or motion of blocks of crustal material. The idea is relatively new and there is no definitive agreement on the subject, even among proponents of the general concept. There may be as many as 20 plates, or as few as six, moving at a rate of between one and five centimeters a year. These plates warp, crack, fold, and fault when they come in contact with each other. It may be a violent collision, such as the meeting of the Indian plate and Asian plate, which occurred with such force that the impact created the Himalayan Mountains. In other places, however, these plates meet in a more gentle manner, one of the plates folding under the other and diving back toward the center of the earth at a steep angle. This downwelling occurs and actually creates the deep-ocean trenches, where the plates are being "consumed" as the geologists say.

What does the future hold in light of plate tectonics? Is a new ocean being formed in Africa? Will California slip into the ocean? The East Africa Rift Valley extends from Mozambique through Ethiopia to the Red Sea and has become six miles wider in the last 20 million years, leading some people to believe the continent is breaking apart and a new ocean is being formed. But this movement is slow, even by geologic standards; in the same 20 million years the Red Sea and Gulf of Aden have spread 200 miles.

California is another story. This area rests on the edges of the Pacific plate, which is moving into the American plate. The action might suck California down with the Pacific plate as it proceeds under the American plate; or the area could split apart from the American plate and become an island in the way Madagascar was separated from Africa. In either case, California appears to have an active future, seismically speaking.

*In the **Red Sea** (opposite page), where plates are diverging, a new ocean may be born.*

***New world.** As crustal plates collide, they are consumed in trenches (below). New mountain ranges are formed adjacent to these trenches.*

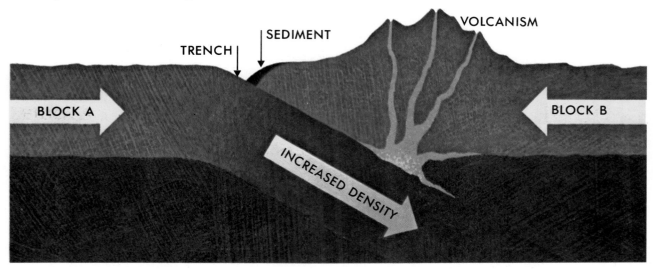

Water Story

Water—The vapor of promise from cosmic furnaces
Water—Growling snarling steam from earth's burning bowels
Water—Born in Hell—Pouring lava into human loins
Water—Deluge in the skies—Rainbows in heavens
    Sea kettles puffing clouds from flowing water snakes
    Frontiers of mind—Glaciers in frozen caps
Water—Wedded to air and earth in dramatic matings
    Carving shores, raising storms, rolling the hopes of man
Water—Injecting salts from land into our thinking blood
Water—Loading quivering gills with gases from the winds
Water—Beating with the moon, heavy in lagoons
    Proud of islands, expecting coral reefs
    Breaking loose furors—Concealing treasures
Water—Magic fluid of the trillion kisses
    Ever returning death of life
    Diluting gold away from greed
Water—Of the original sin
Water—Of redemption
Water—Immense—rare—fragile
Water—Little.

Index

ILLUSTRATIONS AND CHARTS:

Sy Barlowe—18-19; Howard Koslow—13, 23, 28-29, 69, 90-91, 96, 113, 118, 119, 120, 130, 131, 132-133, 138, 139, 141; Howard Koslow and Robert Swanson—14, 15, 33, 52, 128, 134, 135.

PHOTO CREDITS:

AIA—117, 118; Dr. C. Birkeland—71 (bottom); John Boland—71 (top, left); Canadian Government Travel Bureau—42-43; Bob Commer—71 (top, right); W. E. Ferguson—124; Freelance Photographer's Guild: Jerry Chung—111, L. Grigg—76 (bottom), Tom Myers—59 (top), 62; French Government Tourist Office—40-41; Henry Genthe—34-35, 106 (middle), 106 (bottom), 114 (top, left), 115 (bottom); Global Marine—122-123; Tom McHugh, Steinhart Aquarium—64 (top); Jack McKenney—84, 88 (bottom); Richard C. Murphy—56, 58 (top), 58-59 (bottom), 64 (bottom), 73, 89; NASA—21, 47, 57, 68, 99, 125, 140; The Netherlands Consulate General—50-51; Don Nelson—88 (top); Gene Nelson, Hyperion Sewage Plant—86, 102; Chuck Nicklin—70; l'Omnium Francais de Photogrammetrie, Nice—61; Photography Unlimited: Ron Church—100, 101, Rick Grigg—16, 17, 110, 112-113; James Prescott—81; Dr. David Schwimmer—121; Scripps Institute of Oceanography—95; The Sea Library: Ed Angell—2-3, 49; Jack Drafahl—72, Naval Undersea Center, San Diego—67, Chuck Nicklin—48, Elliott Norse—60, Carl Roessler—12, 54; Tom Stack & Associates: Ron Church—24-25, 55, 78-79, 109, Larry C. Moon—32-33, L. Hugh Newman—103, B. S. Oza—127, Eric L. Wheater—142; Surfer Publications, Inc.: Steve Wilkings—31; Taurus Photos: Peter A. Lake—5, Neil Marshall—97 (bottom), 115 (top); United Press International Photo—36 (bottom), 37; Myron Wang—75; Wide World Photos, Inc.—36 (top); © D. P. Wilson—82-83, 85, 87; Don Wobber—65 (top).